# The Righteous Will Flourish
*Living Christian Ethics*

"Steve West and Danielle Gignac have written a clear, logical, accessible and biblical introduction to Christian ethics that can be useful around the globe. Their biblical/theological foundation is followed by seven chapters that deal with just about any issue you'd want to explore: homosexuality, abortion, war, civil disobedience, euthanasia, capital punishment, environmentalism and other timely issues."

**PAUL E. ENGLE**, *USA, adjunct professor and author of* Exploring Worship from Seven Mountains: Delighting in God's Presence

---

"Pointing us clearly to a biblical worldview, Steve West and Danielle Gignac help readers wrestle with the complexities of the ethical issues of our day. They teach us to thoughtfully engage with these important topics while fixing our eyes on God as our source of truth and authority. While different cultures often promote varying perspectives and values, this book will serve as an invaluable resource for the global church, as it directs our gaze to the unchanging character of our eternal God. "

**TIM BEAVIS,** *VP of International Operations, Carey International Pastoral Training, USA*

---

"Someone once said that the Bible is shallow enough for a child not to drown in, yet deep enough for an elephant to swim. *The Righteous Will Flourish* by Steve West and Danielle Gignac is not the Bible. Yet being so full of and firmly rooted in the Word of God, whether you are a freshman, a sophomore or a final year student in the 'University of Christian ethics,' this is a manual you don't want to do without!"

**WEI EN YI**, *pastor, Shalom Church, Singapore*

---

"Steve West and Danielle Gignac's book is scriptural, with good practical pastoral application. The book is very readable and of great value for the evangelical church. I have greatly enjoyed the book and believe it glorifies God and will be a blessing to many."

**THOMAS YIU**, *pastor, Hong Kong*

"In a country like India, where religious plurality is a reality, a Christian is in anguish to distinguish Christian ethics. This book deals with exactly that dilemma. A Christian reading this book will no more be in doubt about being truly ethical no matter what country or culture one comes from. It is just liberating!"

**TIMOTHY BABU**, *pastor, Visakhapatnam, India*

---

"In a time and age where so many ethical issues are on the table, there is a great need for a biblical worldview to deal with each one of them. *The Righteous Will Flourish* is a work highly needed in our days. It gives a solid and biblical foundation for the Christian ethics and how this must be applied to several important issues that every Christian will face in a postmodern society. The present work will surely contribute to the flourishing of the righteous who want to live pleasing their righteous God in the midst of their generation."

**RUBÉN SANCHEZ**, *pastor, Barcelona, Spain*

---

"Christian ethics is not just a philosophical game for brilliant academic minds to play, but it is the framework that shapes daily decisions, wherever we live on the globe. This book gives the strong foundation needed for grounding our ethics in God's Word and then spells out implications, with an awareness of cultural differences that may arise. Moving from complex ideas to real-life situations, the authors help all of us who respond to the name 'Christian' live out our convictions within our own cultural context. After reading Steve's previous book and hearing him speak in Colombia, this newest book again reminds me of my responsibility to think through hard issues and move from the theoretical to the practical—to live out my faith."

**BETH AFANADOR**, *director of El Camino Academy, Bogota, Colombia*

---

"We will be using this excellent, straightforward book as a resource in our college courses on ethics. It portrays grace all throughout, but also our human responsibility as Christians."

**BEADEX WACHIMWA**, *Evangelical Baptist Bible College of Malawi*

# The Righteous Will Flourish
## Living Christian Ethics

**STEVEN D. WEST**
**DANIELLE M. GIGNAC**

GENERAL EDITOR: BOB PENHEAROW

CAREY
PRINTING PRESS

**CAREY**
PRINTING PRESS

*Published by*
Carey Outreach Ministries Inc., Guelph, Ontario, Canada
www.careyoutreach.org

*About us*
Carey Printing Press is the publishing arm of Carey Outreach Ministries, an international Christian organization that provides theological training to spiritual leaders to shape the church and influence the nations.
*General editor:* Bob Penhearow

First published 2019
© 2019 Steven D. West and Danielle M. Gignac. All rights reserved. This book may not be reproduced, in whole or in part, without written permission from the publishers.

*Cover design:* Danielle M. Gignac
*Book design:* Janice Van Eck

*All images:* Shutterstock, public domain or personal archives.
*Bible quotations:* Unless otherwise indicated, all Scripture quotations are from The Holy Bible, New International Version,® NIV® Copyright © 1973, 1978, 1984, 2011 by Biblica, Inc.® Used by permission. All rights reserved worldwide.

ISBN 978-0-9876841-4-1 (paperback)

# Contents

Foreword *by Bob Penhearow* ix
Preface xiii

**PART ONE: FOUNDATIONS FOR ETHICS**
1  Christian theology and the nature of ethics 3
2  Ethics and worldview 27
3  Virtue and ethics in Ephesians 4:1–5:20 53

**PART TWO: ETHICAL ISSUES**
4  Family ethics 81
5  Sexual ethics 109
6  Abortion 135
7  The end of life 159
8  Ethics in society 181
9  Environmental ethics 213
10 Ethics of church leadership 245

# Foreword

The moment Adam rebelled and rejected God's holy and righteous rule in the garden was the moment he and his progeny seized automony. From then on, Adam and his offspring became a law unto themselves with every man doing "what was right in his own eyes" (Judges 21:25). Rather than embracing and submitting to God's ethical rule, men and women continue to actively supress the truth of that rule in order to proclaim and sustain their own personal autonomy (Romans 1:18).

Consequently, down through the centuries man-centred ethics have become the global norm. Man has become his own god and assumed the judgement seat, ruling right from wrong, acceptable from unacceptable and just from unjust. Yet man-centred ethics has notoriously fluctuated, varying from nation to nation and even coast to coast. There are few if any absolute truths as man tenaciously grasps onto his tainted autonomous crown. All in all, our ethical system is in crisis.

In contrast, the gospel transforms man-centred ethics to God-centred ethics. Such ethics reflect a beautiful, holy and righteous God, and reject the tainted crown of autonomy for a joyful embrace of God's holy and righteous rule.

West and Gignac rightly point out that our purpose of being is singular in focus; namely, to bring glory and honour to our triune God:

> We were created to live an ethically righteous life for God the Father, through Jesus Christ, by the Holy Spirit. Everything we do should bring glory to God, and part of glorifying him involves learning to submit to his moral standards with both earnestness and joy. We were made to reflect God's moral character (xiv).

The authors rightly show that there is ethical continuity between the Old and New Testaments:

> The Decalogue was certainly not the whole of God's law, but it did represent the *heart* of it. It showed in summary form how to love both God and our neighbour. As a result, there is continuity between Old and New Testament ethics. It is a mistake to think that the Old Testament ethic was based on legalism but the New Testament ethic was based on love—God's laws were always grounded in love (13).

As an avid traveller, I am often confronted with a variety of ethical issues from spiritual leaders around the world. I realized that leaders today are bombarded by such dilemmas without much structural guidance from a biblical perspective. As such, I approached Dr. Steve West to write a God-centred book on ethics. Dr. West's God-given insights and pastor's heart make him uniquely suited for this authorship.

Together with Danielle Gignac, he created this book to help believers navigate through complex ethical challenges from a glorious God-centred worldview rather than an inglorious man-centred one.

Both authors remain rooted in God's Word as they tackle some of the most pressing issues of the twenty-first century: broken family units, homosexuality, abortion, euthanasia, war, civil disobedience, capital punishment, environmentalism and fragmented church leadership.

I trust you will enjoy the insights of this book and that it will help to equip us all to live a holy and righteous life before a glorious and beautiful triune God.

*Bob Penhearow, D.Min.*
*General editor and president*
*Carey Outreach Ministries*

# *Preface*

> The righteous will flourish like a palm tree,
>   they will grow like a cedar of Lebanon;
> planted in the house of the Lord,
>   they will flourish in the courts of our God.
> They will still bear fruit in old age,
>   they will stay fresh and green,
> proclaiming, "The Lord is upright;
>   he is my Rock, and there is no wickedness in him."
> (Psalm 92:12–15)

*The righteous will flourish.* God's people flourish when they do what is right and honour him. They are fruitful. This does not mean that they will be materially rich—in fact, the Old Testament prophets reveal that many of the righteous were poor. Furthermore, many of the righteous were weak and persecuted. Some of the most righteous people in the land were put to death, precisely because they continued to do what was right in God's sight. Christ's followers in the New Testament likewise experienced persecution, beatings, imprisonment, loss and even martyrdom. Yet, *the righteous will flourish*. God has designed us to live according to his design, and we can only flourish and bear fruit when we walk along the path he has marked for us. To bear fruit, we must be planted in the garden of the Lord. We must be rooted in him, the giver and sustainer of life. Ultimately, we can be righteous because of his character and empowerment. The Lord is upright, and there is no wickedness in him. The fruit that God produces in the life of the righteous is spiritual and moral. He makes us to be the people we ought to be, doing the things we ought to do. It is only in this way that we flourish.

We were created to live an ethically righteous life for God the Father, through Jesus Christ, by the Holy Spirit. Everything we do should bring glory to God, and part of glorifying him involves learning to submit to his moral standards with both earnestness and joy. We were made to reflect God's moral character. As fallen sinners, we need to learn how to live in a way which is pleasing and honouring to him—sin bends our natures so that we want to go our own way rather than God's way. The Spirit guides God's children in accordance with God's revelation, so that we can imitate Christ. Being conformed to the image of Christ is what spiritual flourishing is all about. Since Christ always did what was right, if we are following in his footsteps we also must do what is right. Knowing the difference between right and wrong, and good and evil, is essential for ethics. But it goes further. We must not only know what's right, we must do what's right. Ethics requires accurate knowledge and pure motives that translate into proper actions. There is always spiritual and eternal blessing when we do what is right in the sight of God—even if it results in physical pain or the loss of our lives. *The righteous will flourish.*

This book is an introduction to the subject of Christian ethics. It is published through Carey Outreach Ministries, an organization that exists to train pastors and church leaders around the world. As a result, we have tried to bear in mind that there will be a very diverse audience of readers from many different cultures, nations, backgrounds and languages. Although we have tried not to focus only on Western issues, we cannot help writing from our perspective as Canadian evangelicals. Our goal has been to write about basic ethical issues that affect people all over the world. We have tried to be sensitive to the fact that technological, social, political and economic factors vary from country to country and region to region. We also recognize that not every culture or society around the globe is debating the same ethical issues. In some countries abortion is illegal; in others it is considered a fundamental right. In some countries homosexuals can be imprisoned or executed; in Canada, the government recognizes homosexual marriage and allows homosexuals to adopt children. Depending on which country you live in, a public discussion about homosexuality and ethics would sound very different. Although every country operates with a moral code, there is not universal agreement across the globe about every single ethical issue. One of our goals has been to avoid being too culturally specific, even though we cannot do so perfectly.

This book is divided into two parts. Part One lays down biblical, theological and philosophical foundations for Christian ethics. Part Two is where we evaluate and discuss particular ethical issues. If any reader finds certain sections in Part One difficult, or too theoretical, they can simply move on to our chapters on specific issues. Although the first part is vital to our understanding of ethics from a Christian perspective, Chapters 4–10 can be read profitably even if Chapters 1–3 have not been mastered. Alternatively, some readers may find it easier to read about particular issues and then go back to see the wider foundation on which our understanding has been built.

There are many different ways the material in this book could have been arranged. For example, the issue of abortion can be discussed under the topic of sexual ethics, but it could also be discussed under medical ethics. Often one ethical topic will cross into numerous disciplines and will need to be considered from social, economic, technological, environmental and legal perspectives (to name just a few!). Whole books have been written on every topic that we will be discussing. We are well aware of how short some of our discussions are, especially considering the high importance of the topics covered. Although we would defend our conclusions at length, we have tried to stay focused on being succinct and clear. We do not offer this work as the final word on any of the issues we discuss; it is merely an entryway into the field of Christian ethics.

For us, writing this book was a labour of love. It was work, but it was work that we thought well worth doing. It is our sincere hope that in

God's grace what we have written is true, and that it glorifies and honours him. We also hope that it will encourage people to live in a way that pleases and honours the Lord. May God give us grace to know, to love and to do.

*Steven D. West*
*Danielle M. Gignac*

# PART ONE
# Foundations for ethics

# 1

# Christian theology and the nature of ethics

## ETHICS AND THEOLOGY

As Christians, we believe that ethics and morality are rooted in God. Very simply, ethics is the study of how to make morally right decisions and how to live a life that is characterized by moral excellence. Ethics involves using moral principles and reasoning to determine what we *ought* to do and what we *ought not* to do. Since goodness and moral beauty are grounded in, and flow from, the nature and character of God, if we are going to engage in ethics, we need to understand some basic theology. Ethics—from a Christian perspective—is sometimes referred to as "moral theology." This is an insightful term. Christians do not approach ethics as a subject of abstract philosophy. On the contrary, we approach it as a theological topic that cannot be detached from the existence and character of God, particularly as God's character is revealed perfectly in Christ through the power of the Holy Spirit.

Having a sound theological basis for our ethics gives us a foundation on which we can build a consistent, moral life. Christian ethics, then, will be unashamedly theological. Working out a theoretical ethical system, or making practical ethical decisions cannot be detached from the character of God and his revelation. Although we must use logic, rational thinking and philosophical analysis in our moral reasoning, we must not attempt to do so in an autonomous manner. We need to consciously relate our views of right and wrong to the nature of God: He is our standard. An essential question for ethics, then, is: *How do we know God, and how do we know what God wants us to be and to do?*

**ethics**
*n. the study of how to make morally right decisions and how to live a life that is characterized by moral excellence.*

Moral obligations arise in personal contexts. A chair does not have a moral obligation to a table, but people have obligations to each other. If the chair belongs to someone else, then we have a moral responsibility not to steal it or intentionally damage it. The obligation does not exist to the chair, but to the chair's owner. This means that in order for binding moral obligations to exist, persons must exist in interpersonal relationships. In order for *absolute* moral obligations to exist, therefore, we must exist in a context where we are related to an absolute and ultimate person. This is the only framework in which absolute binding moral responsibility is coherent. John Frame articulates this view admirably when he writes,

> The absolute moral standard must be an absolute person. The Bible is unique in teaching that the supreme moral authority is an absolute person. Other religions and philosophies proclaim absolutes, but those absolutes are not personal. Still other worldviews, like polytheism, teach the existence of supernatural persons, but these are not absolute. But if morality must be based on one who is both personal and absolute, then the God of the Bible is the only viable candidate.[1]

---

[1] John M. Frame, *The Doctrine of the Christian Life: A Theology of Lordship* (Phillipsburg: P & R, 2008), 63.

Since everything we are and do is related to God, morality is inescapable. We are always under the constraints of absolute moral obligation because we live in the universe that was created and is sustained by an absolute and ultimate moral being.

Since God is holy and righteous, he is the standard of all moral goodness. Goodness flows out of his character and being (Exodus 34:6–7). He is pure, just, morally beautiful and harmonious. God is incapable of sinning or doing anything morally wrong. He cannot be wicked. Since God is omniscient, he cannot be mistaken in any of his understanding. Therefore, God never thinks that something is right when it is actually wrong. As a result, the Lord is never confused about ethical issues or uncertain about what ought to be done in any circumstance. We cannot even begin to imagine what it is like to exist this way! To be perfectly good, to have infinite understanding and to have fathomless wisdom is a mode of existence and being that goes beyond our ability to comprehend. We are simply so different from God in our own moral goodness, knowledge and wisdom, that we are not able to imagine how blessed it is to *be* God. What we do know of God, however, should fill us with awe and should drive us to be as conformed as possible to his moral character. We should endeavour to grow more and more like him, and we should strive to put away from our hearts and lives everything that is contradictory to his nature. This means that we must resist sin in our lives, and we must pursue holiness. Part of this pursuit will involve learning to live the way that God wants us to live. We must learn to do things God's way; we must learn to act in conformity with his moral will.

> **We live in the universe that was created and is sustained by an absolute and ultimate moral being—God.**

Every day of our lives we make countless moral decisions to either obey or disobey God. Often, these decisions will be made with a clear choice of either submitting to God in obedience or rebelling against him in sin. For example, we know that it is wrong to gossip—when we are confronted with the opportunity to engage in gossip, we either accept or reject God's commands. Many other times we don't *consciously* make decisions about moral issues—we simply *act* in accordance with the moral character that we have formed over time. Virtually every time we talk with someone, we have the opportunity to tell the truth or to lie. Most of us don't consciously weigh our options when we're talking with people (hopefully we're so used to telling the truth that that's simply what we do!). Likewise, we have the opportunity in our daily interactions to be either kind or cruel, loving or spiteful, encouraging or discouraging, etc. We rarely—if ever—debate whether we should be mean to someone; when we are unkind, it is because we simply are; we act in accordance with the character that we have developed.

There are other times, however, when we find ourselves genuinely puzzled about what is the right thing to do in a given situation. We may consciously think, meditate and debate within ourselves, weigh options, take various factors into consideration and even seek counsel from others. Sometimes situations and issues can be enormously complex, and there seem to be competing pros and cons on all sides. People that we respect may look at the same issue and arrive at different conclusions. We need to recognize that thinking about moral issues can be difficult and confusing, and sometimes making conscientious ethical decisions can be very challenging for us.

**...sometimes we don't want to think about ethical issues, since we may be forced to acknowledge that we need to make some changes in our thinking and acting.**

We also need to honestly admit that sometimes we don't want to think about ethical issues, since we may be forced to acknowledge that we need to make some changes in our thinking and acting. In fact, many times when people start thinking through an ethical issue, they begin to see problems and moral inconsistencies with their own decision-making and lifestyles. For example, if people really think about how they spend their money, or treat the environment, or care for their neighbours, they may find that they are failing to live up to God's standards for their lives. Rather than making the effort to change and pursue moral excellence, it can seem much easier to give up and to try to ignore the issue instead of continuing to think about it. Settling for moral mediocrity and the ethical status quo seems to be a default for many.

In the next chapter we will outline the contours of a basic Christian worldview, and we will consider the relationship between the existence of God and the existence of morality. For now, it will be enough to say that God is the Creator of the universe and that he has woven morality into the deep fabric of reality. God has created human beings in his image (Genesis 1:26), and this includes an inherent sense of his basic moral will and law. The fact that a moral sense is deeply engrained in us is proven by an examination of peoples and cultures across the world and down through history. It is true that some cultures have taken different views on certain issues (such as slavery, war, polygamy, etc.), but there has never been a single culture in human history that has been morally anarchic (i.e., completely without laws or a moral code)—such a society would implode in self-destruction.

**God has woven morality into the deep fabric of reality.**

It is also worth reminding ourselves of the fact that cultures share an extremely large amount of moral common ground—the points of difference have been relatively small compared to the points of similarity. Every culture in the world today, and every culture in history, has believed that there is a profound difference between right and wrong in the moral realm, especially when they believe they have been wronged. People can passionately disagree sometimes about what is the right thing to do—which is why we still have debates about morality—but they are committed to the principle that there is a real difference between right and wrong. There has never been a culture or society where there have not been any moral values and taboos. This phenomenon is best explained by the theological truth that human beings have a moral constitution that is designed by God.

If God has created human beings as moral creatures, why do we find so much human immorality? In order to understand the moral landscape, we need to understand the theological concept of *sin*. People are the image-bearers of God, but they are also corrupted and depraved as a result of the Fall. They resist God's goodness and light. Our consciences are damaged because of sin, and they do not function as perfect guides to right and wrong. Unfortunately, our emotions can mislead us, since sometimes we desire to do things that are wrong. We are masters at rationalizing our sin and finding justifications and excuses for our immoral conduct. Sinners often do what's wrong even when they know it's wrong. Living ethically is not merely a matter of knowledge—it is also a matter of the heart and will.

Thankfully, in his great grace, God has not only given us an inherent sense of his moral law, he has *spoken*. Christians can sometimes forget that God was under no obligation to speak to his rebellious creatures, giving them clear teaching and instructions. Yet, in compassion and love, God has given us his word in the Bible. In theology, we say that Scripture is both necessary and sufficient for God's purposes in special revelation. Only in the Scriptures do we find certain clear and specific truths concerning God's will and character, who we are, God's plan of redemption and how we should live in this world. Even though the Bible doesn't discuss everything, it contains all that we need in order to please and glorify God in our lives. In terms of ethics, then, this means that we must study Scripture in order to find out what God wants us to be and do. If ethics is the study of moral principles and moral decision making, we must not neglect the Bible, since that is where God most clearly reveals his righteous nature. It is also—and this is critical—the place where he gives us moral instructions and commands. We do not need to debate whether something is right or wrong if God has explicitly told us! God is our moral authority, and his Word is our standard, because it is the Word of the living God.[2]

> **The Bible is the place where God gives us moral instructions and commands.**

Christian ethics are grounded in the fountain of moral goodness which is God himself. But Christianity insists that God has revealed himself to us most clearly in the person of his Son, the Lord Jesus Christ. Christ is our authority, teacher and model. We are to put his teachings into practice and also to imitate his example. For example, when we learn about love, we listen to the words of Jesus and also look at his life and sacrificial death. When we wonder about the goodness of creation, we look to the fact that nothing was made except by the Son, he continually sustains all things by his powerful word and his resurrection sanctifies and glorifies the physical realm (starting with his own physical body). Christian ethics are Christ-centred.

---

[2] This is not the place to discuss proper biblical interpretation. Suffice it to say that we do need to practice good exegesis and be sensitive to context. Biblical theology and the progressive nature of God's revelation—which culminates in Christ—needs to be carefully observed. There are biblical reasons why we do not directly practice every single command in Leviticus. Those issues, however, will not be explored here.

## THE TRINITY AND ETHICS

Ethics, being established in God, is deeply *relational*. God is holy and he is love—in fact, he is holy love. According to God's revelation, he exists as a Trinity: one nature shared by three persons. The Father, Son and Holy Spirit are all equal participants in the one divine nature. We distinguish them not in terms of their being and nature but in terms of their personal relationships with one another. For example, the Father relates eternally to the Son as the Father, and the Son relates eternally to the Father as the Son. Their essence is identical in every possible way, so the only way to differentiate them is by *how* they relate to each other in their intra-Trinitarian community relations. The main point of what we are saying is this: God has eternally existed in relationship as Father, Son and Spirit. God's love, then, flows and animates his own being in personal communion and mutual joy among the divine persons of the God who is three-in-one. Simply put, love is meant to be shared in community, and this is something that God experiences perfectly and eternally in himself.

> **Love is meant to be shared in community, and this is something that God experiences perfectly and eternally in himself.**

We want to be extremely reverent at this point, but it is worth asking how the members of the Trinity treat one another. Can we imagine the members of the Trinity failing to love one another? Is it possible to think of the Son swearing at the Spirit in a fit of rage? Could the Spirit go to war against the Father? Would it ever be possible that the Son would lie to the Father, or steal from him or exploit him? Could the triune God ever be internally divided by sin, immorality or unethical behaviour? The answers to these questions are surely obvious.

Although we are not God and we do not exist in Trinitarian form, we are to imitate God in terms of ethical conduct. We have both positive and negative obligations. Positively, we should do things that God delights in. Negatively, we should not do things that God deems immoral. If it would be scandalous for the Son to slander the Father to the Spirit, we ought not to slander others. The Trinity provides a helpful analogy for us to work out ethical conduct in our relationships.

On the positive side, as imitators of God, we are to act in accordance with his righteous moral character. Since God is truth, we are to be honest and refrain from lying. If God does not exploit people or take advantage of them, then neither should we. If God cares for the poor, the oppressed and the marginalized, then we know that these are things we ought to do as well. If God values the sanctity of human life (because he himself is the sanctifier of human life and the ultimate judge of value), then we should reject any ethical position that undermines or denies the sanctity of human life. Leviticus 19:16 states, "Do not do anything that endangers your neighbour's life." In Christian ethics, this means that we must be willing to say to our neighbour, "I'll keep you safe." God is a God of kindness, compassion and love, and we are to cultivate these virtues in our lives in imitation of the Lord. Since our actions flow from our characters, the more we grow in godliness and Christ-likeness, the more ethical our conduct will become.

## LOVE IS ESSENTIAL

Love is an absolute nonnegotiable for ethics. Jesus famously declared: "A new command I give you: Love one another. As I have loved you, so you must love one another. By this everyone will know that you are my disciples, if you love one another" (John 13:34–35). Notice that the standard for our love is nothing less than the standard of Jesus himself. We are to love one another the way that he has loved us. And how has he loved us? How has he demonstrated the depths of his love? There is perhaps no finer passage to answer these questions than 1 John 4:7–21:

> Dear friends, let us love one another, for love comes from God. Everyone who loves has been born of God and knows God. Whoever does not love does not know God, because God is love. This is how God showed his love among us: He sent his one and only Son into the world that we might live through him. This is love: not that we loved God, but that he loved us and sent his Son as an atoning sacrifice for our sins. Dear friends, since God so loved us, we also ought to love one another. No one has ever seen God; but if we love one another, God lives in us and his love is made complete in us.
> 
> This is how we know that we live in him and he in us: He has given us of his Spirit. And we have seen and testify that the Father has sent his Son to be the Savior of the world. If anyone

acknowledges that Jesus is the Son of God, God lives in them and they in God. And so we know and rely on the love God has for us.

God is love. Whoever lives in love lives in God, and God in them. This is how love is made complete among us so that we will have confidence on the day of judgment: In this world we are like Jesus. There is no fear in love. But perfect love drives out fear, because fear has to do with punishment. The one who fears is not made perfect in love.

We love because he first loved us. Whoever claims to love God yet hates a brother or sister is a liar. For whoever does not love their brother and sister, whom they have seen, cannot love God, whom they have not seen. And he has given us this command: Anyone who loves God must also love their brother and sister.

There is enough in this passage to cause us to praise God for eternity! God shows his love by sending his Son to provide atonement for our sins. It is *impossible* for us to be more loved! God is infinite, eternal love, and he has set this incomprehensible love upon us. Christ died for us. Allow that to sink in; don't move too quickly out of that thought. (It is, after all, a thought that will fill our minds throughout eternity.) We are saved and brought into the current of God's love, by his love. Now, having received the gift of infinite love—having been filled with infinite love—we are to pour out love to others. We are to imitate Christ. We are to love.

Since God is a Trinity, he has eternally existed as an internal community of loving persons. Nevertheless, God created creatures outside of himself that he also loves. Since human beings are not triune, we imitate God's love by directing our love to beings that are external to us. God loves people, and we must love people too. In fact, we must not only love our friends and neighbours, we must love our *enemies* as well. Love must be directed toward everyone.

What this means in practical terms is that we must work to enter community, strengthen community, build community and sometimes even create community. If we are to exist as members of a loving community, we will need to live ethically. Love, care and affection must be combined with moral knowledge for such a community to be sustaining and God-honouring. Since we are sinners, we need grace, the Spirit and God's Word for guidance. One way that love will express itself is by motivating us to work for a just and fair society. Love will protect. Love will refuse to settle for the expression of sentimental emotions—it will labour to provide practical good. Love is not weak; it is the strongest force in the world.

## LIVING IN COMMUNITY

It is vital to recognize that, in Christ, God is creating a new covenant community. His old covenant people of Israel were a national body, but many members of that community did not know the Lord in a saving way and died in their sins. Every person in the new covenant community is a believer, however, and knows the Lord and has experienced the forgiveness of sins.[3]

People who live in countries that prize individualism more than communal relationships may fail to appreciate how much the New Testament emphasizes the corporate nature of the church. Jesus saves individuals, but these individuals are part of his body, the church, which is the new covenant community. God's plan is not for a separated and isolated number of individuals to bring him glory—God desires individuals to be brought together, reconciled to each other and reconciled to himself, all through Christ. Community, therefore, is the matrix in which we are to live and love. We cannot please God if we withdraw from human society and try to live apart from engagement with others.

> **Community, therefore, is the matrix in which we are to live and love.**

Love is expressed in community through relationships, and we need to recognize that we live in *communities* where we have different obligations and responsibilities. We live in a web of relationships, some of which are more demanding than others. Our primary relationship is the one that we have with God. We have an absolute ethical responsibility to honour him properly. There is nothing more evil than failing to love God. But we are also to love our neighbour. These two ethical obligations, of course, are the greatest commands in Scripture. Matthew 22:37–40 says,

---

[3] Jeremiah 31:31–34 is explicitly about the new covenant. It is quoted in Hebrews 8:8–12. There are many other biblical passages that are about the new covenant, even when the term "new covenant" is not necessarily used.

Jesus replied: "'Love the Lord your God with all your heart and with all your soul and with all your mind.' This is the first and greatest commandment. And the second is like it: 'Love your neighbor as yourself.' All the Law and the Prophets hang on these two commandments."

Failing to love God and love our neighbour, then, is a most serious sin; it is the height of immorality; it is the greatest ethical failure. There can be no genuine morality apart from loving God and loving others.

Love is also the foundation of the Decalogue (i.e., the Ten Commandments; Exodus 20:1–17), because it is concerned with our relationship to God and to our neighbours. It starts with our vertical relationship with God and then moves to our horizontal relationships with others. The latter are based on the former. The Decalogue was certainly not the whole of God's law, but it did represent the *heart* of it. It showed in summary form how to love both God and our neighbour. As a result, there is continuity between Old and New Testament ethics. It is a mistake to think that the Old Testament ethic was based on legalism but the New Testament ethic was based on love—God's laws were always grounded in love. His laws guided Israel's ethics, and Israel's ethics informed the ethical teaching of Jesus and the apostles. Jesus went beyond the confines of the Old Testament law by bringing new revelation and also by fulfilling the law, but he never rejected God's moral will as revealed in the Old Testament. The heart of both Testaments is that we are to do all that we can to love God and to love and care for our neighbours.

> **It is a mistake to think that the Old Testament ethic was based on legalism but the New Testament ethic was based on love.**

Nevertheless, there are degrees of responsibility that we have for others. On a very broad level, we have ethical responsibilities to the world at large. We have different responsibilities to our nation, however, than we have to other nations. The government of our nation commands our obedience in a way that a government in another country does not. Citizens in Uganda are not under a moral obligation to submit to the government of France. So, although we are to submit to our own government, this general principle will be worked out in a variety of particular situations. Similarly, not everyone lives in the same city, town or region, and there are particular responsibilities that attend living in certain areas. There are also unique obligations that arise from living in proximity to particular people. For example, if a person who lives next door to us falls and hurts themselves, it is our responsibility to help them, and not the responsibility of a person who lives miles away.

Family relationships entail special degrees of interconnectedness and obligation. Those with ailing parents have particular responsibilities to care for them. If we have children, we have an obligation to care for them (which includes an enormous amount of time and resources). A parent's level of responsibility and care for their own children simply cannot be transferred equally to all of the children in the world. Examples could be multiplied, but the point is that we all exist in a special web of relationships, where our level of responsibilities range from very general to very specific. Christians are to take all of their ethical obligations seriously, but there is a hierarchy of moral responsibility and duty in every one of our lives. In the final analysis, however, we must always love.

**We all exist in a special web of relationships, where our level of responsibilities range from very general to very specific.**

A common delusion is to think that we can sin in private ways that will not affect others. We must remember that every ethical decision we make is made inside of a web of relationships—who we are in private will affect others, even if only indirectly. Literally nothing affects ourselves *only*. At one level, this is obvious when we consider that everything we are and do is directly related to God. Even my thoughts are *coram Deo* (i.e., before the face of God). But everything we *think* also shapes our character, even if it is just in small ways. If we are routinely engaging in selfish thoughts, we are not going to be capable of acting selflessly. If we allow ourselves to be rude and spiteful to one person, that will simply make it easier for us to be rude and spiteful to others. Virtue, then, is not and cannot be merely a personal matter. Our moral characters are always in the process of being shaped either positively or negatively.

## POSITIVE, NOT MERELY CONDEMNING

Too often we tend to think of ethical issues in terms of negatives (i.e., thinking about what is *unethical*), rather than thinking creatively about how we can bless others. This ensures shallowness in our ethical thinking and responses. We can believe that abortion is a great evil, but we also need to recognize that we need a healthy and holistic community response to the problem of abortion that actively *helps* rather than merely condemns. If we live in a country that allows for the practice of active euthanasia, we may speak out against the practice, but we should also work to produce societal structures of loving care. Having love and good intentions is *necessary*, but it's not *sufficient*—we also need proper order and structure so that we can effectively minister to people's needs. As Wendell Berry has wisely and rightly said, "Good work is done by knowing how and by love."[4] We need to love, but we also need to know how to help others. Christians are sometimes seen as angry individuals who only condemn others for their perceived sins and moral failures; we must do better. Part of our moral responsibility is most certainly to speak out against evil, but that does not go far enough. Christian ethics is not simply a negative exercise in condemnation. Done properly, it is a beautiful and loving, positive exercise in helping, honouring and blessing others. Christ did not come simply to judge; he came to redeem, heal and save.

> **Having love and good intentions is necessary, but it's not sufficient.**

Although environmental ethics is treated as a special topic later in this book, we will make a comment or two about it now. Sometimes things that are obvious escape us; the fact that we live in an environment can be one of them. God created the world, fish, birds and animals before he created people. Yes, human beings represent the pinnacle of his creative work in this world, and human beings are utterly unique as the image-bearers of God. But the first commands that God gave to Adam and Eve involved being fruitful and multiplying, and exercising stewardship in the world (Genesis 1:28). How we

---

[4] Wendell Berry, *The Art of Loading Brush: New Agrarian Writings* (Berkeley: Counterpoint, 2017), 71.

treat the world God created is of enormous moral consequence (to say nothing of practical consequence). Christians may feel morally responsible to feed the homeless—but do we care about the pollution in the air that the homeless are living in? We may give a cup of cold water to the thirsty—but have we helped them if we've poisoned the water with our waste? Some of us may send money to other countries for famine relief—but if our regular economic practices exploit overseas workers or ruin their environment by ravaging the land of their natural resources, are we truly acting morally? We need to look for positive ethics in the big picture of the environment. Stewardship involves actively helping people flourish, rather than just avoiding harming them. Ethical decisions only take place in the world that God has created, and the world environment must not be ignored in the decisions that we make.

> **Stewardship involves actively helping people flourish, rather than just avoiding harming them.**

## ETHICAL SYSTEMS AND PRINCIPLES FOR MORAL REASONING

We believe that Christianity encompasses the best insights and thoughts of the best ethical systems that have been developed throughout history. In fact, we believe that every valid ethical principle is ultimately based on, and derived from, Christian truth. Down through the ages, great thinkers and philosophers have thought about the nature of morality. Some of them have attempted to come up with systems of reasoning that can generate proper ethical conclusions. We do not have the time or space—or desire—to try to set these systems out in any great detail, but it is worthwhile to briefly describe some of them and point out their most important features. In our opinion, each system tends to make some fair points, but only in Christianity can a comprehensive and sound system of ethics be constructed. Some of the principles in the following ethical systems can be combined into a greater whole. As Christians, we believe that everything we do ought to be done for the glory of God. Yet that general principle of the highest good does not tell us exactly what action will result in God's greatest glory. Having studied the Scriptures, we seek to learn God's wisdom, and then we seek to apply it. Sometimes, however, the Bible doesn't speak directly to the issue that confronts us. What principles and axioms of moral reasoning should we follow in those cases? What insights from ethical systems are reliable? Note that we are not replacing the Bible with philosophical ethics—we are seeking wisdom that works under the authority of God and his revelation.

## 1. Deontological ethics

This approach to ethics derives its name from the Greek word *deon*, which refers to duties, obligations, principles or rules. Deontological ethics maintains that certain things are intrinsically right or wrong, regardless of context or consequences. Morality is always objective and obligatory. A proponent of deontological ethics will not evaluate the potential costs or consequences of an act; they will try to follow unchanging moral principles. Relativism and subjectivism are ruled out. For example, if lying is wrong, then a deontologist will deny that there is ever a time when lying is right (even if lying might save someone's life). In this view, our job is not to evaluate ethical situations and decide if we ought to obey a moral law or precept—our job is to perform our duty, and that means following the moral law.

> **Deontological ethics maintains that certain things are intrinsically right or wrong, regardless of context or consequences.**

Even a cursory reading of Scripture will reveal that human beings are under many moral obligations, and that God reveals many binding moral principles and duties. Biblical ethics are not relativistic or subjective. Because of the nature and character of God, there are certain things that are always right, and there are other things that are always wrong. Hating your enemy is always morally wrong; love is always right. This is not just because hatred generates negative consequences, and love generates positive ones. Love is intrinsically *right*; hatred is intrinsically *wrong*. Murder is never right in any circumstance. Torturing the innocent is likewise always wrong, even if it could produce good consequences. (One possible way that torture could produce good consequences is discussed below.) Even though deontology is an important part of Christian ethics, Christian ethics are more, not less, than deontological. There are other factors and realities that inform a multi-faceted Christian view of morality. However, apart from Christianity, the whole idea of binding moral rules and obligations makes very little sense, but we will explore this more in the next chapter.

## 2. Consequence-based ethics

This is a family of ethical systems that evaluates actions on the basis of their consequences. In this view, actions aren't necessarily intrinsically good or bad—they can only be assessed as good or bad on the basis of the consequences that they produce. The most common type of consequentialism is utilitarianism. In utilitarianism, the governing idea is that something is morally good if it produces the greatest happiness or the greatest good for the greatest number of people.

> **The governing idea of utilitarianism is that something is morally good if it produces the greatest happiness or the greatest good for the greatest number of people.**

Some utilitarians hold to rule-utilitarianism, where the goal is to find the ethical rules that are most conducive to producing good. For example, if people refraining from stealing will produce the best overall consequences in society, then nobody should steal, regardless of their circumstances. Since not stealing is a good moral principle that tends to the greatest good, we should follow it without making exceptions. We are good at self-justification, so it is better to conform to proven moral rules than to rely on our own subjective judgement.

Act-utilitarianism, on the other hand, focuses not so much on rules as on actions: every individual act is evaluated on its own for the consequences that it produces. In this approach, stealing may or may not be morally acceptable—it all depends on the consequences that the act of stealing generates. An act-utilitarian may think that it is acceptable to steal food in certain circumstances, since feeding the starving may produce more happiness than allowing them to suffer.

Although it is important to examine the consequences of our actions, utilitarianism as a full ethical system is not sustainable. First, there is the practical problem of trying to determine what will actually bring about the greatest good for the greatest number. Is building a factory and providing people with jobs a good thing? It would seem so. But what if that factory ruins an ecosystem and destroys a way of life for a future generation? How can we know all of the consequences of our actions? What is the timeframe? It seems that we are simply not in a position to know many of the things that are relevant in a utilitarian model.

Second, is the greatest good a matter of quantity or quality? If three people experience mediocre happiness, is that more important than one person experiencing exquisite happiness? Is the avoidance of pain worth more than the experience of pleasure?

Third, if the ends justify the means, then it seems that many things that we intuitively feel are wrong can be justified. Imagine that watching a horrific example of torture would be very effective in deterring and eliminating crime if people believed they would also be tortured if they broke the law. On utilitarian grounds, one could conclude that society would be better off with much less crime, and therefore, torturing someone publicly was the right thing to do. The individual being tortured would suffer, but the overall good of society would outweigh their pain. Even if the person was innocent of any criminal activity, torturing them could be justified on the principle of the greatest good for the greatest number. But torturing the innocent does seem morally problematic, to say the least.

Fourth, utilitarianism can make it very difficult to protect minorities. If 99% of a population are very happy keeping the remaining 1% as slaves (to the point where the happiness of the majority was greater than the misery of the minority), does that make slavery morally acceptable in their context?

Fifth, what rationale can the utilitarian give for why we should care about the good of others? If the reason is because it's simply the right thing to do, we are back to deontology. Utilitarian considerations may be valid, but they need to be established on a stronger moral foundation than what is provided in the system itself.

The Bible often does call people to consider the consequences of their decisions and actions. Countless passages in the prophets call sinners to examine where their sin is leading them. The consequences of both wisdom and folly are clearly spelled out in the Book of Proverbs. Jesus routinely told people to pay attention to the consequences of their decisions, both now and for eternity. Ethical conduct does generally produce a preponderance of good consequences, because that's how God designed both us and the world. But the Bible does not tell us that a rule or action is right or wrong only *because* of the consequences that it generates. Sometimes doing a morally right thing will result (even predictably) in suffering. It is typically considered heroic if someone willingly sacrifices their own life to save another's, even if the person saved can contribute less to society than the one lost. It was right to save the person regardless of the outcome. Thankfully, rules and actions that are deontologically good often do produce good consequences, even in this fallen world.

## *3. Virtue ethics*

This is perhaps more of a *perspective* on ethics than a full ethical system. Virtue ethics is concerned with the production of a virtuous character, since a virtuous character will produce a life that is marked by ethical flourishing and increasing virtue. There is an unbreakable connection between our moral character and what we do. Actions can form or deform our characters, and our characters generate our actions. There is a reciprocal causality at play here. In virtue ethics, we need to choose to do the things that strengthen moral character; we need to cultivate virtue. Acting from moral principles, putting others first, choosing to love, etc., are all important for the formation of virtue. We need to be virtuous to act virtuously, and acting virtuously makes us virtuous. From a Christian perspective, there is much wisdom, truth and beauty in the virtue ethics tradition. Still, virtue ethics can be practiced by those outside of the Christian tradition, and our definition and understanding of what constitutes virtue needs to be carefully placed *inside* of a Christian framework. We are too sinful to make ourselves virtuous apart from God's grace. We need forgiveness, healing, cleansing and mercy. We need regeneration and the Spirit. The goal is not to have a generically good character; the goal is to have the good character of Christ formed in us. It is also vital to allow the Bible to give us a definition and description of genuine virtue. In our sin, there are times when we can confuse virtue and vice. We need God's transcendent view of what virtue is, and we need his supernatural help to grow in it.

## *4. Divine command ethics*

Some theists have held the view that something is right simply because God commands it. There is something noble in this position, but it requires modification. God only commands things that are in keeping with his character, which is perfectly holy, just and good. He cannot be evil or arbitrary, and he does not change his mind. Sometimes people attack the divine command theory of ethics by saying that it means that God could command dismembering infants, or genocide, or stealing and lying, and then all of those things would be ethically obligatory and morally good. Or, instead of God commanding us to love him, he could command us to hate him (and our neighbour), and all of a sudden hating God would be morally good. This, however, seems to be nonsense. It is better to recognize that anything that God *does* command is morally obligatory, but God will *only* command things that are in accord with his perfect character. God is not a philosophical abstraction; he is a being with a definite character and defined existence. As a result, it is not the commands of God *per se* that are our highest authority: it is the *God* who issues the commands. He stands by his words, and it is precisely because his commands issue from his holy and righteous character that they are always good and morally obligatory. Divine commands are always binding, because God himself is our standard. He is eternally, immutably and perfectly good. God is love. The loving commands of the God who is love can be nothing less than morally binding.

**God stands by his words, and it is precisely because his commands issue from his holy and righteous character that they are always good and morally obligatory.**

## CHRISTIAN ETHICS ARE MULTIFACETED

From a Christian perspective, then, we can recognize something of value in each of these four approaches to ethics. Many elements in them should be combined; Christian ethics are indeed multifaceted. When we weigh our moral options, we need to consider the deontological status of the action (i.e., if God has spoken clearly about it, or if God's view of the action can be logically inferred with surety, then we know what we ought or ought not to do). Doing the right thing, however, can be done for the wrong reasons (which, from God's perspective, destroys any moral praiseworthiness of what we do). For example, Jesus rebuked the Pharisees because they were saying their prayers publicly to be seen by people and be noticed by them (Matthew 6:5–8).

Praying is a good thing to do, but if we abuse it with sinful motives, we are acting immorally. The same is true of giving to charity, serving in the church, preaching a sermon or writing a book. We can twist almost anything that is good into something evil through our sinful motivation for self-glory or gain. This is one of the reasons why the virtue ethics tradition is so important.

> **We can twist almost anything that is good into something evil through our sinful motivation for self-glory or gain.**

An evaluation of the possible or likely consequences of an act is not a sufficient basis for a full ethical system, but this does not mean that consequences are irrelevant in our ethical decision making. Let's say there have been severe famines in two places, and we want to provide money for hunger relief. If food costs twice as much in Country A as it does in Country B, our donation could feed twice as many people in Country B. If there are no overriding factors, it seems reasonable to donate money to the place where it will be used to feed the most people (which produces the greatest good for the greatest number). Utilitarian concerns are legitimate concerns, but they need to be placed in a better framework. It is when we come to the Bible that we find the foundation on which all proper ethical intuitions and insights can be grounded and built.

Christian ethics, therefore, are not simplistic; there are multiple levels of analysis in moral decision-making and behaviour. Furthermore, God's evaluation of our morality does not stop at the level of the external act. We are in complete agreement with the following analysis:

> Moral assessment is something we all do routinely, whether we know it or not, and includes evaluating the action, the motive behind it, the consequences of the action, and the character of the person doing it.[5]

---

[5] Scott B. Rae, *Introducing Christian Ethics: A Short Guide to Making Moral Decisions* (Grand Rapids: Zondervan, 2016), 18. This is an excellent, concise introductory resource. It is well worth reading.

When we evaluate moral options and decisions, we need to bear in mind these different elements. We also need to remember that people may judge us on the basis of our external acts, but God is also aware of the thoughts and motivations of our hearts. Things that may impress people may be abominable in the sight of God. Conversely, things that people revile may be honourable in God's eyes. Doing something good for the wrong reason is actually immoral. Good intentions, however, are not enough either. Many people do things out of sincerity, but nevertheless what they do ends up harming others. Given our own sin and the difficulty that accompanies trying to live ethically and help others in this fallen world, we must recognize that we are utterly dependent on God for grace, wisdom and moral goodness. Cultivating a close walk with God is essential for ethical living.

> **Cultivating a close walk with God is essential for ethical living.**

## FRUIT OF THE SPIRIT

We will close this chapter with a few observations about the fruit of the Spirit. Many Christians have memorized Galatians 5:22–23, but may not have noted the context. The fruit of the Spirit begins with the word "but," which indicates a logical connection to the preceding material. Although a consideration of the proper context would include an examination of the whole letter, we will only back up to verse 19. Here is the passage:

> The acts of the flesh are obvious: sexual immorality, impurity and debauchery; idolatry and witchcraft; hatred, discord, jealousy, fits of rage, selfish ambition, dissensions, factions and envy; drunkenness, orgies, and the like. I warn you, as I did before, that those who live like this will not inherit the kingdom of God.
>
> But the fruit of the Spirit is love, joy, peace, forbearance, kindness, goodness, faithfulness, gentleness and self-control. Against such things there is no law (Galatians 5:19–23).

Notice the strong contrast between the fruit of the flesh and the fruit of the Spirit. There are numerous things that could be said, but we only have space to identify a few that are most relevant. First, our sinful nature is the opposite of God's character. When the Spirit is producing fruit in our lives, what he is really doing is making us like Christ, who is the Son of God incarnate. If you want perfect examples of the virtues that are included in the fruit of the Spirit, read the Gospels and examine the life of Jesus. Second, the works of the flesh are internally destructive. People who are characterized by these things are divided and fractured; they lack wholesomeness and internal integrity. In contrast, when the Spirit produces his fruit in our lives, we gain internal health. Our hearts and minds can rest because they are brought into alignment with the way that God has designed us to be. Third, the fruit of the flesh *destroys* relationships and community, while the fruit of the Spirit *builds* relationships and community. This last point is essential for ethics.

Go slowly through the list of the works of the flesh. Do you want to live in a society that is characterized by these things? Do you want to be surrounded by friends who are filled with hatred? Do you want to be married to someone who is crassly selfish or is routinely flying into a rage? Is living in a world characterized by sexual exploitation really the environment in which we want to raise children? The answers to these questions are too obvious to be stated. But, *but*, look at the fruit of the Spirit. What about living in relationships characterized by love? What if there was patience rather than rage? Peace rather than war? What would it be like if all of our relationships were characterized by faithfulness? There is simply no comparison between the ways of living depicted in this text. The only way for anyone to grow deeply moral is to walk in fellowship with the Spirit who is our guide and counsellor. If we want to live ethically and be a blessing to the world, we need to cultivate a vibrant relationship with the living God, through Jesus Christ, by the Holy Spirit.

## REFLECTION QUESTIONS

1. What are ways that your church can help create a more loving community, both inside of itself and in the wider society?

2. How does the fact that God is a Trinity help us understand ethics?

3. What are some biblical examples of good actions that were rejected by God because the person's heart motive was wrong?

4. In your own words, summarize the four approaches to ethics found in this chapter. What examples can you think of in your own life or experience where these types of ethical reasoning were followed?

# 2

# Ethics and worldview

If we are going to have a deeper understanding of ethics, we need to spend some time thinking about worldviews.[1] Our worldview is the large-scale framework in which we understand reality. It is the way we *view the world* and interpret it. Everyone has a worldview, even though many people have not consciously analyzed theirs, or are even aware they have one. One main reason many people disagree about particular ethical issues is because they have a much deeper and more fundamental disagreement about their respective worldviews. As we will see, ethical decisions are not made in a vacuum—they are dependent on a large number of factors and other beliefs.

## FOUNDATIONS OF A WORLDVIEW

In order to have a coherent and intelligent ethic, we need to learn about worldviews—our own included—and how to analyze, articulate

---

[1] The material in this chapter is adapted from Steven D. West, *Head, Heart, Hands: Life-Transforming Apologetics* (Guelph: Carey Printing Press, 2015), 21–61.

> **worldview**
> *n.* large-scale framework for how to interpret the big picture.

and evaluate them. There are a variety of schemes that scholars use to analyze worldviews, but the major points are always more or less the same. The following discussion, therefore, is fairly standard. For our purposes we will break down worldview foundations into five key categories. Being aware of one's worldview for ethical decision-making and understanding morality is very important. Understanding our own worldview—as well as the worldviews of others—can be critical for finding our way forward when we seem mired in ethical debates or disagreements.

### 1. God

Belief or non-belief in the existence of God is a fundamental point of division between worldviews. Atheists and Christians cannot have the same worldview. These are not, however, the only two possibilities.

> *Belief or non-belief in the existence of God is a fundamental point of division between worldviews.*

For example, pantheists believe that God is everything and everything is God. An atheist cannot be a pantheist, but neither can a Muslim. Islam holds to an absolute monadic Allah; Christians hold to a Trinity. For real worldview-level agreement, it is not merely a matter of believing in the existence of a supernatural being: what you believe about the nature of God matters tremendously.

Absolute agreement on every miniscule theological detail, however, is not required in order to share a broad worldview commitment. At the most basic level every Christian shares the same worldview, even though Christians will have a variety of opinions about smaller issues. Some Christians believe that God exists outside of time and others believe that God exists inside of time (and some believe that he existed outside of time prior to creation and exists inside of time after creation). Some Christians believe that God wants believing parents to baptize their infants, while others insist baptism is only for those who have made their own decision to follow Christ. It is critical to understand, however, that these are *internal* debates that take place *inside* the larger Christian worldview. Some issues are more fundamental than others: training yourself to discern what's most important is essential. A diversity of perspectives on secondary issues can exist when people are building on the same primary foundations.

## 2. *Metaphysics*

This second foundational block focuses on the essence of reality. It is concerned with the true nature of things—what something actually is, or a thing's *is-ness*. (This subject is also called *ontology*.) When we think about God and metaphysics, we see how our first and second worldview building blocks overlap. In philosophy God is often treated as one metaphysical (or ontological) object like the rest, since you can ask questions about his nature and existence, as you can for other things like whales and stars. In worldview terms, however, the existence and nature of God is so important it is best to treat it as a special category. God's existence and nature is also profoundly different from everything else. For example, God is self-existent, self-sustaining, and he depends on nothing outside of himself. Everything in the entire physical universe is completely dependent on him. Thus, even if God is thought of in metaphysical terms, it must be with the recognition that his metaphysical status is categorically unique.

Key worldview-level metaphysical issues are contained in questions like the following: What type of existence do abstract concepts and ideas have? Do mathematical truths exist apart from knowing minds? Is the physical universe all there is? How do we differentiate one object from another? What is the essential nature of a particular entity? Metaphysical issues can seem rather abstract (and some are), but others are both practical and vital. In our contemporary times, we find strict materialists who believe that matter is all there is, we find animists who believe that some animals, trees, stones, etc. have spirits, and we find some Hindus who believe that what appears to be the material universe doesn't exist at all—it's an illusion. These are substantial differences. So, although metaphysical discussions can get very deep and very abstract, how we answer certain metaphysical questions is an essential part of how we understand and interpret reality.

## 3. *Human beings*

In one type of analysis, human beings could be considered under the heading of metaphysics. But, as with God, the way human nature is understood creates a significant divide between worldviews. What makes us human? Materialists believe that, like everything else, human beings are composed exclusively of matter. People do not have souls or spirits. For most materialists, human beings emerged as a result of the unguided and goalless process of biological evolution. Our existence in the universe was neither foreseen nor designed. We are a cosmic accident without any overarching reason for our existence. When a person dies their physical body simply decomposes. There is no life or consciousness after death.

For those who hold to reincarnation, however, life does not end at death, nor does it begin at birth. People are thought to be stuck in a brutal cycle of birth, death and rebirth. Only during a minority of time will a soul even be attached to a human body (which really does raise significant questions concerning human nature in this model). Although in the Western world some people seem to regard the idea of reincarnation as fantastical (after all, who doesn't want to swim in the ocean as a dolphin or soar over the mountains as an eagle?), for those who live in the Eastern world, the concept of reincarnation is agonizing. The whole goal of life is to eventually *escape* from reincarnation's horrible and painful cycle. Whether someone believes reincarnation is good or bad is, for our purposes, irrelevant: what is important is that those who believe in reincarnation and those who don't have irreconcilable views of human nature. Christians believe that people have one physical birth and death, after which they are judged by God. This belief entails that human beings are more than physical bodies. In the end, the Christian view, the materialist view, and the reincarnation view are so different that their adherents cannot share the same worldview. How a person understands the nature of human beings is, therefore, a core component of their worldview.

### *4. Morality and ethics*
When we are considering worldviews, we must remember that people do not need to agree on every particular ethical issue in order for them to share the same worldview. Nor do they need to share worldviews in order to reach the same conclusions on certain ethical issues. Some Christians and atheists share a common commitment to pacifism, whereas other Christians and atheists believe that there are circumstances in which it is morally acceptable to wage war. Some secular

humanists believe that pornography is harmless; others believe it is dehumanizing and exploitive. There are some particular issues, however, that do reveal fundamental worldview differences. For example, Christians believe that the highest ethical duty in the world is to glorify God and love him supremely. Needless to say, atheists are not able to agree.

In worldview terms, we need to think about the nature of morality itself, more than about individual ethical issues. How does a worldview influence views on morality? One big difference would be whether morality is objective or subjective. Is there a universal standard for right and wrong that applies to everyone at all times? Are all moral standards simply community guidelines that are, in the final analysis, arbitrary and non-binding? Is morality discovered in the commands of God in the Bible, or are those commands to be evaluated and judged on the basis of the ethical sensibilities of the human reader? Does morality even exist in the universe, or are ethical judgements really just expressions of our particular emotional preferences and tastes?

> **In worldview terms, we need to think about the nature of morality itself, more than about individual ethical issues.**

There is a lot more to morality and ethics than debating contentious issues. Some people deny that morality is universally binding, whereas others reject the whole concept of morality. The former group holds that all morality is relative, while the latter group believes that the very idea of genuine morality is mistaken. In stark contrast, Christians believe that morality is objectively real. They maintain that God has given ethical decrees that are to be obeyed and moral principles that are to be followed. Applying ethical principles is challenging, and there are tough cases where even people who share the same broad worldview come to different conclusions about what should be done in particular cases.

It is also the case that over time many people come to refine their previous views, sometimes even coming to very different conclusions than they held originally. This ethical development can take place inside of the same, basic worldview framework. It can be very helpful to acknowledge the diversity of ethical opinions that exist within a shared worldview, and how ethical opinions can evolve without necessarily overturning the entire worldview structure. Worldviews can be separated on the basis of meta-ethics (i.e., the overarching nature of ethics), so this should be the focus of worldview discussions. To be a Christian, one does not need to be a pacifist, nor does one need to be an advocate of just war theory. We need to learn to be able to discern when diverse opinions can be accommodated in a shared worldview. Since this is the case, we should demarcate sub-issues that are irrelevant to whether or not one is a Christian.

### 5. Epistemology

In the language of philosophy, the subject that deals with knowledge is called *epistemology*. What can be known? Is there truth? Are some methods of investigation better than others for discovering truth? What is the difference between *knowing* something and *believing* something? Even if we know certain things, can we know that we know them? How can we be intellectually and rationally justified in holding the beliefs we have? Are all intellectual views merely a matter of relativistic perspective, entailing that all beliefs are equally true or false? Should we be radical skeptics? Is what we call "truth" really just a disguise for maintaining convenient beliefs to secure power and control?

**Key elements of a worldview**

How does a worldview consider the following:

1. God
2. Metaphysics
3. Human beings
4. Morality and ethics
5. Epistemology (knowledge)

It is not much of an overstatement to suggest that contemporary philosophy is obsessed with epistemological issues. The technical debates are extraordinarily complex. Most of the people with whom you discuss your faith, however, are not going to be specialists in this field! You do not need to master the subject to be able to identify a few key points. For example, sometimes you will hear someone asserting that you cannot know anything for sure. This is surely a self-defeating opinion: How could they *know* that? There are others who maintain that the only things we can know for sure are learned through the scientific method. For them, human knowledge is legitimate when it is scientifically obtained, but all other knowledge claims (eg. about morality or God) are illegitimate. Nobody who holds this view, however, discovered it scientifically, so it also refutes itself! It is a philosophical rather than a scientific opinion. On the other side of the spectrum, some insist that humans were designed to know God, and it is through his revelation that we can have confidence in our knowledge. Relativism, scientism and the Christian view are obviously mutually exclusive.

People may agree that we can know things, but disagree about *how* we know them. Or people may disagree about certain facts, but agree on the best method for obtaining knowledge. Human knowledge cannot be considered apart from human nature, metaphysics, and the existence of God. Some thinkers want to add that human knowledge cannot be severed from ethics as well. They argue that certain virtues (eg. honesty and humility) are required for us to think properly. Everyone has biases and prejudices, and we can use our minds to justify our false beliefs, rather than use them to discover the truth. Being virtuous and being intelligent are not the same, but virtue is not irrelevant for thinking and analysis.

## THE CHRISTIAN WORLDVIEW

At this time we will provide a very brief sketch of what the Christian worldview says about each of the five key elements. We want to be able to show people that we have a grasp of ethical issues and principles, but also that our morality is rooted in a coherent worldview system. We may even want to try to demonstrate that the Christian worldview is the only worldview in which ethical reasoning ultimately makes sense. We are never ashamed of God's ethical standards: this is God's world, we are his creatures, and morals and ethics are at home in Christianity. Lots more could be said, but for now, we will turn our attention to sketching out the contours of the Christian worldview in terms of God, metaphysics, human beings, ethics and epistemology.

### 1. God

In regard to the first major element, perhaps it can go without saying that Christians believe in the existence of God. It is desperately important, however, to make sure that people have a rough understanding of what that means. The Christian God is not a generic supreme being who created the world and then stands idly by watching history unfold. Neither is he a benignly benevolent, grandfatherlike deity who just wants us to play nice in our earthly nursery before bringing us to live with him in heaven. The God revealed in Scripture is mighty and majestic, sovereign and perfect in power, knowledge, wisdom and being. He is moral flawlessness and radiant righteous goodness. Although he is transcendent and high above us, he also chooses to enter into intimate personal relationships with his people.

There is no one like God. He is the only being that is self-existent, self-sufficient and dependent on nothing. God is holy. He is spirit. He is triune, which means that God exists with one nature that is shared by three interpenetrating persons—absolute unity in diversity, the One and the Many. God the Father, God the Son and God the Holy Spirit can be distinguished on the basis of their personal relationships with one another, but they share the same essence. There is even more to God than we can imagine or will ever know. He is infinite, and therefore we can never exhaustively fathom his nature. Nevertheless, God has revealed himself to us in the person of his Son, the Lord Jesus Christ. Christ is indispensable for our knowledge of God. It is in Christ that God is most fully revealed and comprehended by human beings.

> **Christ is indispensable for our knowledge of God.**

### 2. Metaphysics

Since only God is self-existent, it follows that the physical universe ultimately depends on him for its existence. In terms of metaphysics, then, God is the independent creator of the universe. He envisioned it in his mind and brought it into existence by the power of his will. Matter does not exist eternally or independently. Material reality is not the primary level of reality. Before there is matter, there is mind and spirit. The natural laws that regulate the material universe and all physical interactions are designed and established by God. Scientists can study physical interactions inside of this order, but the very existence of this physical reality and the laws by which it operates are dependent on the mind, will and power of God.

> **According to Christianity, the most important thing to understand about the universe is that it is creation.**

A biblically-informed worldview understands that the physical universe is created by God. According to Christianity, the most important thing to understand about the universe is that it is *creation*. It is not eternal, nor an inexplicable fact, nor something that happened to pop into existence one dreary afternoon. No, the universe is creation. This metaphysical system is the handiwork of God. It reveals his power and glory. It is stamped with his imprimatur. Every star and every microbe, every blade of grass and every supernova, from the smallest sub-atomic particle to the universe as a whole, top to bottom, front to back, side to side—it has all been created. On one level, metaphysics is really the subject that deals with the study and exploration of God's creation.

### 3. Human beings

The sheer diversity of created entities is staggering, but one of the most incredible things that God has created is you. Human beings were the climax of his creative work on the earth. So, the third key component of a worldview—human nature—also receives attention in the Bible. Humans are not here on earth as a result of a series of unguided, accidental events. We are not merely biological systems without intrinsic purpose or value. We are not cosmic orphans. Although we do share much in common with the animal kingdom, there is a *qualitative* difference between human beings and all other animal species. God has given human beings the special and unique gift of a living spirit. We are physical *and* spiritual beings.

More than this, we are also called the image-bearers of God. Precisely what this means has been debated, and we won't enter into those waters now. The main idea is not too difficult to grasp. First, God is a spirit, and he has given each of us a spirit. Second, we are self-conscious, intelligent beings with a high capacity for rational reasoning. Third, we have moral sensibilities and a God-given understanding of the difference between right and wrong. We are morally responsible for what we do. These three points naturally function together. A fourth understanding, which is in harmony with these first three points, may have occurred to the original readership of Genesis. In the ancient near east, the language of "image and likeness" could refer to statues bearing a king's likeness and image that were often set up around a kingdom's boundaries. Wherever the king's image was found, it was a sign that the land belonged to him. Follow-

**What makes human beings image-bearers of God?**

1. We have a spirit.
2. We have a capacity for rational reasoning.
3. We are morally responsible.
4. Our existence declares the worldwide kingship of God.

ing this line of thinking, wherever God's image-bearers went, they would be living signs that God was the king who ruled the land. In other words, as humans spread over the world, their very presence would be a proclamation that the world belongs to God. Perhaps all four of these points provide a composite, harmonious glimpse into what it means to bear the image of God. We are created to be rational, responsible, physical-spiritual beings whose very existence announces the worldwide kingship of God. Being created in God's image denotes both ontology and function.

If the most fundamental truth about the universe is that it is creation, the most fundamental truth about human beings is that they are created in the image of God. However, when we look around the world today—to say nothing of looking back through human history—it seems that human beings are capable of performing incredibly cruel acts and producing a breathtaking amount of evil. How can this be, if humans really bear the image of God? How can God's image-bearers be so evil and corrupt?

Sadly, the biblical depiction of the human race is not only one of glorious creation in the image of God. The Bible also records the rebellion of the king's image-bearers against their sovereign. Despite being warned about the consequences of rejecting his will, the first humans decided to disobey God and defiantly go their own way. As a result, they freely forfeited their original closeness with God. Instead of finding a more exhilarating way of living—as they anticipated—they died spiritually and corrupted human nature. Biblical anthropology teaches that humans are created in the image of God, but are currently labouring under the ruinous consequences of choosing to fall into sin and evil. Human beings still bear God's image but it is horribly marred and defaced. Now instead of being characterized by righteous goodness we are too often guilty of self-centredness and wickedness. By nature, human beings are far nobler and far more ignoble than we tend to think.

### 4. Morality and ethics

These considerations about goodness and wickedness link the topic of human nature with the subject of morality and ethics. The Bible takes a definite position on our fourth main worldview issue. Many people who know a little bit about Christianity tend to believe that the Bible contains lists of commands that begin with "thou shalt" or "thou shalt not" (with more of the latter than the former). In other words, most people associate the Bible with moral rules or a code of ethics. Although the Bible is about far, far more than ethics, the Scriptures do contain moral imperatives, and they are deeply concerned with right and wrong. We ought to rely on them to help us navigate through this fallen world. Given the existence of God, morality is assumed to be real and objective. People may not like God's moral standards, but the standards exist nonetheless. The Bible teaches that people everywhere in the world have a general moral sense of what is good and what is evil. Our problem is hardly ever that we don't *know* what's right, it's that we don't *want to do* what's right.

From a Christian perspective, morality is rooted in the nature of God himself. It is an objective part of his character and essence. Evil is whatever fails to conform to God's nature. God knows the difference between good and evil, and he has revealed it to us in his Word. He has also created us with consciences that are generally in tune with the moral law. Unfortunately, we are experts at excusing our behaviour and finding ways to be the exception to the moral rules that bind everyone else. Human moral practice, therefore, is always going to be fundamentally flawed in this world. We have a disposition to evil that education cannot erase. Yet we still know that there is a profound difference between right and wrong. A biblical worldview

**Key elements of a Christian worldview**

1. **God:** is self-existent, self-sufficient and dependent on nothing.
2. **Metaphysics:** God is the independent Creator of the universe.
3. **Human beings:** uniquely possess a spirit and are image-bearers of God; fallen into sin.
4. **Morality and ethics:** moral standards are revealed through Scripture and human conscience.
5. **Epistemology:** people can have a genuine knowledge about God, themselves and the world.

insists that God is perfectly good, he has created a world where morality is important, and human beings really do know there is a categorical difference between right and wrong. The main problem is not that we lack all moral knowledge, it is that we fail to do what we know we should. Yes, there are difficult moral issues that can face us, but most of our moral decisions in daily life are clear, whether we like it or not.

## *5. Epistemology*

Claiming that we have moral knowledge and that we know the difference between right and wrong leads us to epistemology. In order to have moral knowledge, we have to have, at the least, some knowledge in general. God is omniscient: he knows everything that can be known. God created human beings in his image, and at least part of that concept includes our faculty for understanding and our ability to reason. Since God is omniscient, it follows that knowledge is not only *possible*, it is *actual*—God is a God who *knows*. Furthermore, since God knows everything, he knows how to communicate truth. He also knows how to create a world that exists objectively outside of our minds, and how to calibrate our minds and senses to this external world, so that when we interact with it we can learn about it and grow in knowledge. Even more importantly, God knows how to create us so that we can grow to know him better and better. The Bible teaches that God is truth itself and he never lies, so Christians have confidence that God will not deceive them nor create an absurd universe. Because the God of truth designed human nature and the world into which he placed us, we can trust both our senses and minds, and accept the existence of non-relativistic truth.

## IN THE BEGINNING

One of the most fascinating facts about the Scriptures—from a philosophical point of view—is that the five key elements of a worldview are all present in the first two chapters of the very first book of the Bible. As a matter of fact, the very first verse in the Bible gives us our first two major elements: "In the beginning God created the heavens and the earth" (Genesis 1:1). Here we have the presupposition of the existence of God and a declaration about metaphysics: God created the world.

As the first chapter of Genesis unfolds, God speaks and brings a living world into being. There is a cause and effect relationship between his will, his verbal command and the creation and formation of the world. The fact that God can speak and produce what he desires would be impossible without knowledge—God couldn't bring about his creative desires unless he knew how. God also decides to create human beings in his image. Here the categories of God, knowledge and metaphysics are all present, and they intersect in the creation of human beings. As a result, before the conclusion of the Bible's first chapter, we discover four of a worldview's five foundational elements.

It is tempting to argue that the first chapter of Genesis also references morality and ethics because of its use of the word "good," but this is not a sound inference. It is true that God not only creates, he *valuates* (i.e., assigns value to his work), and refers to his creation as "good." The word "good," however, in this context probably does not refer to ethical or moral goodness but to functional or aesthetic goodness. A hammer can be a "good" hammer, but not in a moral sense. God made light and saw that it was good, but the light did not have the property of moral purity or ethical righteousness. So even though we think we can assert that Genesis 1 contains statements of value theory, we don't think we can push too hard for the presence of an explicit morality.

A stronger case for the presence of morality and ethics in the first chapter of the Bible can be made on the basis that Genesis 1:28 contains God's instructions to Adam and Eve to be fruitful and multiply and to be his stewards over the created order. Some take this as the first command in the Scriptures, which means it was morally necessary to obey it. Are these instructions, however, *ethical* instructions? It seems that they are: God is giving his human creatures a job description that is their ethical responsibility to follow. However, the ethical nature of this command becomes more apparent in subsequent Scripture. For argument's sake, we will take the overly skeptical position that ethics are not referred to in Genesis 1.

Nothing is lost even if we only have four out of five key worldview elements in Genesis 1. The fact that there are *four* is downright amazing. Soon after, in Genesis 2:16–17, God does give a clear ethical command to Adam. Thus, within less than two full chapters of the first book of the Bible, all the major categories for the biblical worldview appear in embryonic form. Of course they are not fully articulated yet and they will be developed at great length through the process of God's progressive revelation, but still—so quickly—they are all there. And, although this should go without saying, their presence is not *ad hoc* (i.e., the text was not written with the agenda of articulating worldview foundations). In other words, Genesis was written long before any philosopher ever sat down and thought about worldview systems. The pre-critical, pre-scientific, pre-Enlightenment, pre-technological Book of Genesis—in less than two chapters!—stakes

Amazingly, within the first two chapters of the Bible, the five major areas for the biblical worldview appear in embryonic form.

out a worldview foundation at a time when nobody had ever thought about analyzing worldview foundations. Given that Genesis is concerned with theology rather than philosophy, it is astounding that everything we need for the foundations of a philosophical worldview are present.

Only one chapter later in Genesis, the tragic fall of the human race into sin and the resulting spiritual separation from God is recorded. At this early juncture we are given the key to understanding how human beings are so valuable and capable of greatness, but also so vile and capable of evil. We have become a host of contradictions. The text also records, however, how even in justly punishing his fallen creatures God shows mercy and grace. He clothes them and promises them that one day the deceiving serpent will be destroyed by a descendent of the woman (Genesis 3:15). This promise is often referred to by theologians as the *protoevangelium*, the first gospel, the prototypical promise of redemption and salvation. This prophecy and promise is fulfilled by Jesus Christ; he is the solution to the human problem of sin. The first three chapters of the first book of the Bible, then, provide us with the foundations that are required for a worldview. They help us understand God, the world, and human nature (both the good and the bad). The Bible, of course, goes on to develop these themes at great length, so Christians have an incredible wealth of material from which to construct, interpret, and test their worldview. Since God's Word is true and accurately describes reality, we find that the Christian worldview is intellectually coherent, rationally rigorous, emotionally satisfying and pragmatically workable.

**Protoevangelium** refers to the first gospel promise found in Genesis 3:15:

"And I will put enmity
 between you and the woman,
 and between your offspring
 and hers;
he will crush your head,
 and you will strike his heel."

## GENUINE MORALITY

If we are being honest, we know that our moral judgements are at least as emotional as they are rational. Some people go so far as to argue that our ethical views are literally nothing more than the product of our feelings. They maintain that, in just the same way as we find rotten food disgusting, we find certain behaviours disgusting. These feelings of revulsion are subsequently rationalized and placed in a mental compartment we label *morality*. In this interpretation of morality, the following three statements really amount to the same thing:

1. I believe that murder is immoral.
2. I find murder disgusting.
3. I don't like murder.

It is important to notice that this approach to morality can tell you something about someone's *feelings* about various things, but nothing about the *nature* of the things themselves. We may not like broccoli, but that doesn't show that it has the intrinsic property of being disgusting. We may not like murder, but what does that prove? It does not take a keenly trained logical eye to see that the claims "I don't like murder" and "Murder is evil and immoral" are hardly identical, nor is there a necessary, logical bridge from the former to the latter.

Nevertheless, there is an intuition underlying this interpretation of morality that is extremely important: emotions play an enormous role in our ethical evaluations. But so should our intellects. Moral judgements are too important to be left to the whim of our feelings. In fact, one of the very obvious features of morality is that it cannot be merely based on our emotions. Ethical challenges arise when we know we *ought* to do the opposite of what we *feel* like doing! Animals operate on instinct, but we are to use moral reasoning to overcome instinct. Human beings are not supposed to do whatever they want to do, as we are often influenced by sinful desires. Ethics deals with obligation and duty, regardless of how we feel.

**Ethics deals with obligation and duty, regardless of how we feel.**

Having said this, however, we simply can't suppress our emotional responses to moral realities: how we feel about them can be a necessary part of reaching proper moral conclusions and also confirming our moral sensibilities. For example, cruelty or genocide should not only be intellectually unacceptable to us, they should also make us *feel* disgust, sorrow, outrage and revulsion. Seeing a beautiful act of loving kindness, however, should fill us with feelings of joy and peace. We are not calculators; we cannot determine right and wrong using algorithms and abstract reasoning alone. We are created as emotional beings, and there must be a place for that in our moral universe. Proper moral function will have a place for both intellect *and* heart, emotions and obligations.

When we begin to take this matrix of emotional and intellectual interaction seriously, we discover a chicken-or-the-egg type dilemma.

Aristotle (c.384 B.C.–322 B.C.), considered one of the Fathers of Western Philosophy, believed that when a virtuous person discerns the most virtuous path, they take it; and so, virtue leads to virtue.

Aristotle recognized that a person needed to be virtuous in order to form accurate moral judgements. He believed that when a virtuous person discerns the most virtuous path, they take it. This exercise in virtue confirms their virtuousness and strengthens it. Thus, a virtuous person grows in virtue throughout the course of their lifetime. In this process, virtue is required for virtue, virtue leads to virtue, and virtue is necessary in order to recognize the virtuous course of action. But where do you start if you aren't very virtuous in the first place? (The same type of problem is discussed in the world's great wisdom literature: it takes wisdom to discern wise teaching, and wisdom to practice it. A wise person will grow in wisdom, but how can a fool begin to become wise?)

We will not try to solve this dilemma here. (Hint: the Christian response has a prominent place for grace and God's intervening help.) For now it is sufficient to observe that our views on morality are generated at the intersection of our intellectual and emotional faculties. One of the huge advantages of the Christian worldview is that it takes this intellectual-emotional reality of our human experience seriously. Christianity has a significant place for value and values. Morality, ethics and virtue are all at home in the Christian worldview. They are not considered in a cold, calculating, abstract or dispassionate fashion. On the contrary, in Christian theology, moral values are vibrant and lively. They are infused with colour and beauty. When we witness a particularly moving moral act, we often find ourselves saying, "That was a beautiful thing to see." In fact, beauty itself—and our aesthetic experience of it in art, music, dance, architecture, gardening, literature, etc.—thrives in Christianity. This is worthy of reflection and meditation.

Even the existence of evil, interestingly enough, points to the reality of objective goodness. It is an inescapable reality that the world does contain a great deal of suffering and evil. The counterbalancing truth is that the world is also full of beauty, goodness, value and joy. There is evil, but there is good. In fact, the existence of evil is parasitic on goodness. Evil is a negation. Its existence consists in the entirely negative sphere of deforming and devaluing, the same way that sickness can only exist as a negation of health. All of this assumes that proper forms and values pre-exist their deformations: parasites require hosts. To judge something as evil is to say that it ought not to be; something has gone wrong; something inherently positive has been compromised. Identifying something as evil rests on the tacit assumption that there is good—and the good is something logically prior and deeper. It is foundational and an integral part of the fabric of reality.

## CHRISTIANITY AND VALUE

In contrast to naturalism, Christian theology and morality are organically connected. Morality is not merely compatible with Christianity, it is an integral part of it. God's universe is a moral order by virtue of his nature and his design of creation. There is not merely a moral dimension to the universe; morality is woven into the very fabric of creation. Human beings are moral beings who have moral knowledge and make moral judgements on the basis of an objective moral standard. This standard is not abstract and impersonal, but concrete and personal. The standard of moral goodness is the perfect character of a holy and loving God. In God, fact and value, *is* and *ought*, are merged in an eternally coherent relationship.

Morality is rooted in God's character and flows from him. Since the universe is his creation, and since human beings are his image-bearers, people cannot escape his objective moral demands. God is the standard of goodness, the one who issues moral commands and imperatives, and the one who holds people accountable for the good and evil for which they are responsible. In the Christian worldview, morality is intelligible, objective and undeniable. As we have seen, people cannot cogently deny the existence of moral reality.

Space precludes a detailed discussion of other views, such as polytheism, pantheism or animism, but we will say just a brief word about each. Every worldview needs to have its foundational blocks explained and related to one another in a logically consistent fashion, but all of these views fail that test. Specifically, when it comes to ethics, polytheism has many gods and goddesses, but none of them are absolute, so they cannot be the source of absolute and personal obligation. (Finite gods are also often depicted as doing things that are positively immoral.) Animism is similar—where do the spirits come from? What makes one spirit good and another one evil? Animism does not provide a sufficient explanation for the existence of the world, spirits or morality. In pantheism, since everything is God, those who do evil and those who do good are equally part of God—but then the distinction between good and evil breaks down. How can God do evil? Given pantheism, then, God is both good and evil, but this is not only contradictory, it begs the question as to where the standard and source for good and evil can be found.

These very brief thoughts could be expanded at great length, but they will have to be sufficient for now. It is our hope that Christians will not only be bold to stand for God's moral truths and specific commands, but that they will also courageously and humbly insist that the Christian worldview is the proper home for morality and ethics in the first place. When other positions are examined, they are all found to be insufficient. So, when we turn to an examination of specific ethical issues, we want to bear in mind that the foundation on which ethical issues are established and assessed is grounded in the God of the Bible, his incarnate Son and his revelation.

## CASE STUDY: IS MORALITY COMPATIBLE WITH A MATERIALISTIC, ATHEISTIC WORLDVIEW?

In this case study, we present a philosophical argument that is designed to show that morality and materialistic atheism are incompatible. Although this section can be very helpful for Christians who discuss morality and ethical issues with non-believers, it is not essential for an understanding of Christian ethics. As a result, readers who do not find this material relevant may simply move on to the next chapter.

All worldviews must accommodate the fact that morality is a deeply integrated and foundational part of reality: it is insufficient to claim that morality and immorality are illusory or unreal. Human beings are inescapably moral beings. Ethical relativists are a dime a dozen in the classroom, but nowhere to be found in the real world—those who endorse ethical relativism don't live out their view for more than a minute. Unlivable theories can only be supported theoretically, but not practically. Reality has a way of popping pretentions. People may deny the objectivity of morality, but they can never do so with consistency—either in theory or in practice. We have no choice but to take moral standards and axioms as givens (given by God, declares the Scriptures). A potentially acceptable worldview, therefore, must be able to coherently relate objective moral standards to the metaphysical structure of the universe. This can be illustrated by examining the relationship between materialistic atheism and morality. Attempting to provide an intelligible account of the relationship between morality and metaphysics is one of the weakest points of materialist worldviews.

First, it must be noted that many naturalists recognize that objective morality is incoherent given the assumptions of a materialistic worldview. Materialists posit a universe that came into existence from an absolute nothingness without any governing rationality behind it or purpose for it. With these origins it is virtually impossible to avoid the conclusion that the universe is not originally moral, nor is morality found within it. Nothingness is completely void of moral properties. Recognizing the inherent presence of morality, however, begs the question: Did the moral properties pop into existence out of nothing when the singularity appeared, or were they produced by the force of the Big Bang? Many atheist philosophers have rejected the idea of objective morality and ethics, simply because such things do not fit coherently into their understanding of the nature of the universe. Secular humanists who attempt to hold to both objective morality and materialist atheism are swamped in contradictions, regardless of how self-assured their claims are.

For the sake of the narrative, we will grant that the impossible happened and the universe came into existence out of nothing. Even if this is how the physical universe began, we have discovered nothing about the origins of morality. The universe certainly wasn't a moral system during the first minute of its existence. In fact, the materialist maintains that throughout billions of years, matter moved through the universe, colliding, uniting, dividing, building and destroying. Even today we see matter in motion, where one material entity collides with another and causes great damage. A meteor collides with a moon, or a supernova destroys neighbouring planets. The resulting destruction is catastrophic, yet it would be entirely inappropriate to say that the destruction should be interpreted in moral categories. An asteroid may destroy or be destroyed, but nobody argues that such destruction should be considered in moral terms. We know perfectly well that matter in motion is not morally responsible for what it does, nor for the consequences that stem from its interactions with other material entities. Matter in motion is not moral or immoral, it is amoral—the category of morality simply does not apply in any coherent or rational way.

On the assumptions of materialism—and this is critical—human beings are nothing more than matter in motion. Everything about us reduces to material components. When someone punches someone else, matter in motion collides with another material entity, and damage results. Given materialism, however, there is nothing that makes that interaction a moral interaction, any more than two meteors colliding is a moral interaction. Some atheists grant this, but others insist there really are relevant differences. For example, meteors do not have central nervous systems or the ability to feel pain. Since they can't experience pain, they can't be hurt, and therefore the destruction of a meteor does not violate their moral rights. This is true to a point, but insufficient. A tiger that successfully hunts a deer hurts it, but we don't judge that the tiger has done anything immoral. Even if a tiger becomes a man-eater we don't charge it with immorality. The morality of an event, therefore, cannot simply be chalked up to whether or not the victim experiences pain.

Another suggestion that materialists typically make is that the difference is one of personality, intention and rational thinking. Human beings deliberate about their actions and are responsible for them in a way that other material entities are not. Because people have the cognitive capacities required for purposeful decision-making, and the further ability to know what effects are likely to be produced by their actions, what we do can be judged as good or evil. According to this line of reasoning, the reason that human matter in motion is moral is tied to human intentions and intelligence.

This response is wholly inadequate for a number of reasons. One of the reasons is that it simply doesn't go deep enough in its analysis. Human beings think, plan and act on the basis of intentions. But what generates human thought? For a materialist, human thought is produced by nothing more than the physical brain. Needless to say, the physical brain is nothing more than matter in motion—it is the cobbled-together product of blind evolutionary forces. According to many leading Darwinians, the human brain was shaped through countless positive adaptive mutations that conferred survival advantage on the organisms that were lucky enough to mutate in such positive ways. This physical organ controls the physical organism: it is matter in motion controlling matter in motion. This means that every single thought the brain produces is generated from non-sentient, non-self-conscious, non-intelligent, whirring little atoms: in materialism, all human thinking comes from matter in motion. There is no explanation for human intention in this model.

**In materialism, all human thinking comes from matter in motion.**

How can people be held morally responsible for their conduct, if what they do is nothing more than the result of the matter in motion inside of their head? DNA constructed their brains, and it is very difficult to see how people can be responsible for what their DNA does (or what their parents' DNA does in them, if you will). People have no control over the atoms and electrochemical interactions that constitute their brains—how can they somehow become morally responsible for what these chemical interactions cause their physical bodies to do? The brain regulates our moods and our nervous system and our muscle movements and our bodily functions. If the brain causes our arm to move through space and strike another body, it is sheer category jumping to say that such an interaction of two material bodies is part of the moral domain. As has been often said, given materialism the brain secretes thought like the liver secretes bile. How is secreted thought a proper foundation for morality? In a materialistic universe, there is nothing that can lift human beings out of the purely amoral condition of matter in motion.

On the local level of the human organism, then, the concept of morality is inapplicable, because all human behaviour is nothing more than material movement. A massive entailment of this is that human beings are not—indeed, *cannot* be—responsible for what they do. If all human action arises from physical-chemical-electrical interactions over which we have no conscious control, then all of our emotions, dispositions, ideas, thoughts, intentions, judgements, decisions and actions are things that happen to us and in us. I have no control over whether my brain puts into effect a chain of sequences that results in my punching someone, stealing their things, giving them a gift or risking my life for them. If responsibility and control are required for moral behaviour—which they are—then materialism precludes moral action. We do things, but we do not do things morally, nor do we do moral things—morality simply doesn't apply.

On a broader level, it is also worth noting that the nature of a materialistic universe makes the *very concept* of morality unintelligible. Not only are human beings not moral beings, and not only are we not capable of moral action, morality itself is very problematic in a naturalistic order. Morality, of course, is not comprised of matter. When we ask about "goodness" we are not asking about a particular lump of molecules. Does "goodness" and "evil" emerge only when matter interacts with other matter in certain ways? Some materialists try to find a home for goodness or "the good" as a timeless, unchanging abstraction. But it's difficult to demonstrate persuasively that an abstraction has any type of independent existence apart from minds. How did "goodness" exist when there was nothing? It is one thing to say that "goodness" has somehow just existed eternally as an abstract standard, but this is hardly very consistent with the metaphysics of a naturalistic worldview. It amounts to special pleading. Was being unkind to children really objectively immoral before anything existed? What kind of existence could moral standards have when they were alone in a sea of nothingness?

The types of concerns found in the last paragraph are the subjects of seemingly interminable debate.[2] They cannot be profitably pursued here. What is worth noticing, however, is that materialists have a very real problem even if we grant that goodness has some kind of actual ontological or metaphysical status as a timeless, abstract principle. On a practical level it boggles the imagination to think that good and evil are abstract, eternal moral categories, and then a universe came into existence out of nothing, matter collided with matter, and over time this matter became self-conscious and somehow *figured out the true nature of these immaterial moral standards!* The odds of such a thing are incalculable. To think that fragments of matter would bind together and then get in touch with these abstract standards, especially when matter and morality are deemed to constitute completely separate types of realities, is a narrative that seems to depend on blind faith and incredulity.

Plato (c.427 B.C.–347 B.C.) was an influential Greek philosopher who believed that the soul was trapped in the body and needed to find release back to the eternal realm of the forms.

When we add that the evolutionary process aims at survival no matter what the costs, we have to realize that the process by which matter allegedly attained the ability to identify these timeless moral standards is aimed in *exactly the opposite direction from the content of these moral standards.* Roughly speaking, the evolutionary process is selfish, but morality is selfless (i.e., it calls us to fulfill personal responsibilities even if they entail personal loss). Abstract moral standards are not personal or physical entities—they cannot communicate to people. But neither could matter in motion somehow reach out to them, and find communion with these eternal, immaterial, abstract standards of

---

[2] The great divide between the philosophies of Plato and Aristotle was based on similar issues. Medieval philosophy was also characterized by astonishingly subtle debates on the nature of abstract entities. For those centuries, the nominalists and the realists fiercely disagreed about these matters. Perhaps it is safe to say that contemporary debates are not as rhetorically charged and emotionally heated, but philosophers still tend to divide into the broad camps of either the Platonists or the Aristotelians.

good and evil. So, besides the metaphysical problem, there exists the pragmatic problem: Why would we think that unguided matter in motion *accidentally figured out the true nature of eternal, objective morality*? A materialist universe still fails to explain where these standards came from in the first place. Furthermore, it is absurdly mind-boggling to think that these eternal, abstract standards are exactly the principles that *happen to be most conducive to the continued survival of an evolved species inhabiting a material universe that came into existence out of nothing*. How could evolved brains learn a timeless moral code, and how could that abstract, eternal code just happen to be the single best guide for their physical survival? Either our genes have tricked us into believing that morality is real when it is an illusion, or morality is part of a coherently designed plan for the flourishing of human life.

Since the time of the skeptic David Hume, it has rightly been recognized (by many) that an ethical *ought* cannot be derived from a material *is*. Science is descriptive, not prescriptive: it observes and describes what is. It is not able to say what *ought to be* in an ethical sense. A witness can see and describe a robbery, but they do not directly see the morality of the act. A moral judgement represents how we evaluate an event: it is our response to what we observe, and therefore cannot be part of what we actually observed in the first place. Scientists, therefore, can describe what *is*, but this does not legitimize the move from describing what one observed to passing moral judgements about what should or should not have been observed. (Given materialism all human behaviour is determined, so pontificating about what ought to be done is entirely pointless anyway. In fact, even our moral evaluations are determined, since all of our thoughts are determined by the state of our brains, which in turn are determined by nothing more than the blind, non-sentient and amoral interactions of molecules and energy.) The claims "that caused pain" and "it is wrong to cause unnecessary pain" belong to completely different realms of analysis. Materialists have no way of bringing these spheres together.

Some consistent atheists have accepted the fact that materialism and morality cannot be held together in any consistent fashion. Some have argued that morality is a figment of our imagination, and that all things are equally amoral. This position is anarchic and destroys all moral value. (We suspect it is also intellectually fraudulent and that nobody lives it out.) In the end, this position entails that torturing and molesting orphaned children is morally equivalent to feeding them. Nobody really believes this is true, and nobody lives it out. If someone tries to make a show by insisting they see no moral difference between raping a child and building a hospital, they are worth pitying, but not debating. Morality can be denied verbally, but it is impossible for us to live without it.

Those seeking a refuge in moral relativism or subjectivism—if consistent—will find that their position leads in a straight logical line back to the dead end of moral nihilism (i.e., the view that morality is really nothing, that moral claims are meaningless, and that nothing can be more or less moral than anything else). If moral values and ethical judgements are nothing more than arbitrary community standards, or the subtle workings of will to power, or an illusion foisted on us by our genes—or whatever else—then we don't have objective morality. This will bring us back to moral nihilism, but moral nihilism is profoundly untenable. Yet it is the moral theory that coheres the best with the metaphysics of materialism. A rational analysis reveals that it is irrational to be both a materialist and to hold to the existence of genuine morality.

Some people have attempted to get past this by locating morality in the opinions and judgements of a whole community. They argue that morality is what a particular cultural community decrees. Yet the same considerations outlined above for individuals would also apply to collections of individuals. This view also makes it impossible to correct abuses in a particular culture, since *anything* that happens in that culture must be deemed morally acceptable by definition. If a culture accepts slavery, then that is right for them. If the Nazis perpetrate the Holocaust and involve the world in war, that is right for them. In fact, any attempt at social reform would have to be evil by definition. If cultures are able to make anything right, then whatever the majority accepts is right, and it becomes *wrong* to oppose them. It is a very, very strange—and broken—moral position that says that attempting to end slavery or prevent the Holocaust is morally wrong. Thankfully, in Christianity, there is an appeal to a transcendent God—who is an absolute, moral person—so that the perspectives of individuals, cultures and societies are judged against the unchanging standard of his transcendent and eternal goodness.

## *REFLECTION QUESTIONS*

1. How would you explain to someone the relationship between God and the existence of morality?

2. Think about a friend who has a different worldview than you do. How do they understand morality and ethics? Do you think their view makes sense?

3. Review the five major foundation blocks for a worldview. What do you believe about each category?

4. What passages in the Bible can you think of that relate to the material covered in this chapter?

# 3

# Virtue and ethics in Ephesians 4:1–5:20

In this chapter we examine Ephesians 4:1–5:20, a section of Scripture that contains a good number of practical instructions for holy and ethical living. We have already observed that Christian ethics is multifaceted, and an exposition of this passage reveals that Paul presents holy living as an integration of virtues, attitudes and actions. As we study the following verses, we need to keep our own particular culture and church context in mind. We need to interpret the text properly using good hermeneutics and principles of exegesis, but we also need to understand our own contexts so that we can apply these truths with insight and precision. The reader is encouraged to think about how each one of Paul's points can be illustrated from their own culture and personal experiences.

As we work through this passage together, there are many profitable things that we will have to skip over. This treatment is not intended to be exhaustive; it is merely a brief sketch. One of the first things we will see in this passage is that our ethical conduct is to flow out of theological truths. This is the main reason why we have included this chapter in the book. We need a constant reminder that our moral character and ethical conduct is rooted in the nature of God and our redemptive standing in Christ. Christians must not approach ethical issues only with their own reasoning and opinions. On the contrary, they must ground their reasoning and perspectives in God's Word. This chapter provides an example of how a particular biblical text can be used to explain the connection between theological truths and our ethical conduct. By God's grace and through his Spirit, these truths change our hearts and minds, with the result that our actions are transformed. Since this passage is not merely theoretical, appropriate questions to keep in mind as we proceed are, "Am I like that?" and "Do I do that?" We want to understand the text, and we also want to evaluate our lives on the basis of what God's Word says. Ethics cannot stop with theory—it must be put into practice.

> **Our ethical conduct is to flow out of theological truths.**

## EPHESIANS 4:1

Those who study Ephesians have pointed out that Chapters 1–3 contain many theological and doctrinal statements, and Chapters 4–6 apply these doctrinal truths to the believer's life. The ethical and practical application is built upon a firm theological foundation. (This, of course, is just a rough way of dividing the book—there is both doctrine and application interwoven throughout the whole.)

Paul begins his theological application by writing, "As a prisoner for the Lord, then, I urge you to live a life worthy of the calling you have received." Eager to apply the theological material that he set out in the first three chapters of this epistle, Paul uses the strong language of urging his readers to *change how they live*. Their lives are to reflect the greatness of the calling that they have received. The image that Paul uses is of a typical scale in the marketplace. If something heavy is placed on one side of the scale, the pan sinks down. The only way to bring it back up is to put something equally heavy on the other side. What Paul is saying here is that the way we live ought to be equal to the calling we have received. You put all of the blessings that are yours in Christ on one side of the scale, and then you put your life on the other side of the scale. Paul is saying that they should balance. The way you live your life should show the world the true value of the gospel. Your transformed life should be as weighty as God's call and gift of salvation.

## EPHESIANS 1:3-14

If we are supposed to live a life that is worthy of the calling that we have received, we need to remind ourselves of the nature of that calling. Although the entire first three chapters of Ephesians need to be read in order to grasp the magnitude of what Paul is saying, we only have space to examine one critical sentence. This sentence of Paul's, however, is 202 words in the original Greek, and it runs all the way from 1:3 to 1:14 in our translations! These words are among the most beautiful and profound words that have ever been written:

> Praise be to the God and Father of our Lord Jesus Christ, who has blessed us in the heavenly realms with every spiritual blessing in Christ. For he chose us in him before the creation of the world to be holy and blameless in his sight. In love he predestined us for adoption to sonship through Jesus Christ, in accordance with his pleasure and will— to the praise of his glorious grace, which he has freely given us in the One he loves. In him we have redemption through his blood, the forgiveness of sins, in accordance with the riches of God's grace that he lavished on us. With all wisdom and understanding, he made known to us the mystery of his will according to his good pleasure, which he purposed in Christ, to be put into effect when the times reach their fulfillment—to bring unity to all things in heaven and on earth under Christ.
>
> In him we were also chosen, having been predestined according to the plan of him who works out everything in conformity with the purpose of his will, in order that we, who were the first to put our hope in Christ, might be for the praise of his glory. And you also were included in Christ when you heard the message of truth, the gospel of your salvation. When you believed, you were marked in him with a seal, the promised Holy Spirit, who is a deposit guaranteeing our inheritance until the redemption of those who are God's possession—to the praise of his glory.

Before returning our attention to Ephesians 4, we will simply highlight a few key points from this passage. The first thing that we need to understand about this one sentence that comprises 1:3–14 is that every word of it comes under the heading of "Praise be to the God and Father of our Lord Jesus Christ." Paul is listing reasons for believers to worship God. The Father has blessed his people *in Christ* with every possible spiritual blessing in the universe. We must not miss this very important point: spiritual blessings are *only* found in Christ, but *nothing* is lacking for those who are found in him. Even today while we struggle on earth, we have every blessing of the heavenly realms if we belong to Jesus.

Paul uses the little phrase "in Christ" to communicate some extremely important theological truths. Every believer is united with Christ and shares in his glory. We are saved because we are united with him in his death and resurrection. As a result, we also share in his ascension and reign, so in one sense we are already reigning with him in the heavenly realms. All of Christ's righteousness and merit is ours.

> **Because we are in Christ, we are to act in accordance with his character.**

In Christ, we are heirs of God and look forward to living in the new heaven and new earth. The incredible privileges that belong to us by virtue of our union with Christ also call us to live in a way which is pleasing to him in this world. Because we are in Christ, we are to act in accordance with his character. Our position in him is one of perfect righteousness, so we must try to live as righteously as possible in our daily lives. Believers are in Christ, and Christ is the sum of all moral beauty and perfection—this reality is to shape and drive our ethic. Christ is love and holiness incarnate: those who are in him ought to exemplify love and holiness as well.

One of the blessings that is ours in Christ—and every single one is a reason to praise God—is that we have been chosen for the purpose of being holy and blameless (v. 4). When God looks at us, his all-knowing and righteous eyes see the perfect righteousness of Christ, the one with whom we are united. He chose us because of his love, and because of his love he determined that ours was to be the destiny of adopted sons and heirs (v. 5). This choice was based *entirely* on his own love and grace. It was not based on our own merit, but rather on the merit of Christ. It was a decision of pure, free and sovereign grace (v. 6). This was all in accordance with God's pleasure and will—it was simply what God wanted to do. It was all accomplished through Jesus, so that God's glorious grace would be praised eternally. These blessings are far more valuable than all of the world, and they are given as free gifts of grace to God's chosen ones. The value of the gift is beyond all calculation and appreciation. It is infinite. It is inexpressible. It is ours so that we can praise the God and Father of our Lord Jesus Christ.

Paul is just getting started. He goes on to declare that in Christ we are liberated and bought out of slavery; we are redeemed. The redemption price for our freedom is nothing less than the blood of Christ (v. 7). Through this shed blood—through this atonement—there is the forgiveness of sins. One must feel something of the weight of that blessing. *The forgiveness of sins.* That which separates us from God, that which necessitates death, that which mars and ruins and kills—it is atoned for. The Father forgives. Our one great problem is sin, and God in Christ, out of the riches of his grace, has provided for our forgiveness. We do not work to earn it; we *cannot* work to earn it. Our sin demands our life, our death and our blood, but Christ died and shed his blood in our place. In him there is forgiveness of sins, through his blood. His grace is lavishly and abundantly poured out upon us.

**The redemption price for our freedom is nothing less than the blood of Christ.**

God, in his infinite wisdom and understanding, pours out upon us both grace and insight. He teaches us the mystery of his will, the purpose of which is inseparably bound up with Christ. This is something that we could not know until after the incarnation, cross, resurrection and ascension. It is something that we could not understand until the Spirit came at Pentecost. Paul himself could not figure it out apart from special revelation. The mystery of God's will is that everything in heaven and on earth finds its point and significance in Jesus. In the future consummation, all things will be seen as integrated in Christ; he is the centre of it all. There is no meaning for anything apart from its relationship to him. Since we are in Christ, the entailments for this are well worth pondering.

Paul reiterates in 1:11 that we are chosen in Christ, and that God has predestined us according to his own plan. God is the one whose purpose and will ultimately determine everything that happens. God's will for those who are in Christ is that they exist for the praise of his glory. To ensure that God's will is accomplished and that his glory is praised,

Paul says that God chooses his people and predestines them in Christ.

Beyond this, they are also sealed with the Holy Spirit of God (v. 13). The Spirit does many things, but in this text he functions as the *guarantee* that God's children will persevere until the consummation of their salvation (v. 14). God has chosen redeemed sinners to be his very own unique and special possession. And it's all for the praise of his glory.

We could spend a lot more time thinking through the significance of this passage and expounding its riches, but we need to remain focused on how this applies in Ephesians 4:1–5:20. (The reader needs to bear in mind that the rest of Ephesians must be considered as well for full impact.) For our purposes, this is what we need to do: Take a moment to reflect on the weight and significance of God's plan, the gospel of Jesus, the gift of the Spirit and all of the blessings that are ours in Christ. All of these realities stand behind the calling that we have received. Do our lives look like we've been given such a glorious calling? Do the people who know us get a sense of the majesty and weightiness of the gospel? When Paul turns to application, he begins by urging every believer to live a life worthy of the calling that they have received. Our ethical conduct, therefore, is one way that we show the weight and worth of the gospel.

## EPHESIANS 4:2–16

If we are to measure up to this calling, how should we live our lives? What sorts of things characterize a life that is worthy of the calling of Christ? Perhaps we might immediately start to think about stories we have heard of brave Christians who sacrificed their goods or even their lives for the sake of Christ and the gospel. Or perhaps you have suffered great loss for your testimony of Jesus, or you know believers who have. It is entirely possible that your church is filled with people who have suffered persecution for their faith. Regardless of your personal experience, or the experience of the people that you know, Paul does not say a life worthy of God's calling is necessarily a life of great and obvious sacrifice. We might expect Paul to mention something that would be perceived as enormously attention-getting, but he actually says something quite different. Read the following words carefully:

> Be completely humble and gentle; be patient, bearing with one another in love. Make every effort to keep the unity of the Spirit

through the bond of peace. There is one body and one Spirit, just as you were called to one hope when you were called; one Lord, one faith, one baptism; one God and Father of all, who is over all and through all and in all.

But to each one of us grace has been given as Christ apportioned it. This is why it says:

"When he ascended on high,
    he took many captives
    and gave gifts to his people."

(What does "he ascended" mean except that he also descended to the lower, earthly regions? He who descended is the very one who ascended higher than all the heavens, in order to fill the whole universe.) So Christ himself gave the apostles, the prophets, the evangelists, the pastors and teachers, to equip his people for works of service, so that the body of Christ may be built up until we all reach unity in the faith and in the knowledge of the Son of God and become mature, attaining to the whole measure of the fullness of Christ.

Then we will no longer be infants, tossed back and forth by the waves, and blown here and there by every wind of teaching and by the cunning and craftiness of people in their deceitful scheming. Instead, speaking the truth in love, we will grow to become in every respect the mature body of him who is the head, that is, Christ. From him the whole body, joined and held together by every supporting ligament, grows and builds itself up in love, as each part does its work.

The very first thing that Paul says when he teaches us how to live a life worthy of the calling that we have received is that we are to be *humble*. A worthy life is characterized by certain attitudes. It is concerned first and foremost with personal relationships. It is not so much what we *do*, as much as it is who and what we *are* that balances the scales. This is one of the lessons about love that we ought to learn from 1 Corinthians 13. It is possible to know everything, speak eloquently, give away all that we possess and even die as a martyr, but if we do any of this without love, it was all meaningless and useless. Love is not reducible to an emotion, but it is also not reducible to actions. The God-honouring life is not first characterized by a great display of heroism—it is characterized by humility.

> **The God-honouring life is not first characterized by a great display of heroism—it is characterized by humility.**

It is also characterized by *gentleness*. And *patience*. And *bearing with one another in love*. All of this makes sense, and it all goes together. When we look at what God has done for us, and we recognize that it is nothing but a gift of pure grace, where is pride? It has to be excluded entirely! Entering the kingdom requires a gift of grace so that we can recognize how poor in spirit we really are. If receiving the grace of God in Christ is the diametric opposite of human accomplishments and pride, then we simply must be humble. How do humble people act? Are they harsh with one other? Do they dominate or push others aside? No. They put other people first. They serve without being servile. They aren't concerned with themselves or if they "appear" humble. Humble people are too busy with the needs of others to worry about whether their humility is being recognized and acknowledged. Neither humility nor gentleness are to be equated with weakness. In fact, they are both the products of quiet strength and kindness. If you picture a very strong man tenderly cradling a helpless infant in his arms, you have a picture of gentleness. Moral strength is required for both real humility and genuine gentleness.

Those who are humble and gentle will be patient (or long-suffering). Patience refers to the ability to endure exasperation and provocation without losing self-control. It is the opposite of quickly getting angry and lashing out. Those who are patient are able to be controlled and spiritually calm.

> **patience** *n.* the ability to endure exasperation and provocation without losing self-control.

They understand God's grace in their lives, and they are able to extend that grace to others. This allows them to "bear with one another in love." Paul's realism here is refreshing and helpful. You can only bear with people who are bothering you! We do not bear with people who are sources of joy and delight to us. So the fact that we are told to bear with one another (and in this context they are other believers), is proof that in a fallen world there are going to be times when Christians hurt and annoy each other. Someone who understands the calling that they have received will be able to persist in love with their brothers and sisters in these circumstances. As the recipient of God's love and grace, the humble, gentle and patient believer will demonstrate the value of their calling by extending God's love to others. If God has loved sinners like us, how can we fail to love those who sin against us or who conflict with us in other ways?

Paul argues in Ephesians 4:3–6 that all of those who are in Christ are bound together in an incredible *unity*. There is only one Lord. Read the verses again and see the emphasis on the word "one." This unity and oneness is all God's doing. The Spirit *creates* the unity, and yet we are exhorted to make every effort to *maintain* it. Our ethical responsibility is to preserve what the Spirit has made. If we phrase it negatively, we are not to do anything that hurts or destroys the Spirit's creation. By the Spirit we are bound together in peace. This unity and bond of peace must not be desecrated by our attitudes or actions. The Spirit unites; we must not divide. In fact, we must allow the bond to tighten, so that we can be drawn ever closer together as a community of peace. This may not be easy, which is why we are told that we need to *make every effort*—we need to strive to maintain what the Spirit has created. It is far easier to destroy something than to maintain balance or build with care.

> ...we are not to do anything that hurts or destroys the Spirit's creation. By the Spirit we are bound together in peace.

Verses 7–16 shift from a focus on unity to a focus on *diversity*. In 1 Corinthians 12, Paul used the metaphor of a body to make the point that the church only functions properly because it contains an array of diverse parts that are brought together in a fundamental unity. In this passage, Paul is teaching that Christ distributes gifts and also determines the qualitative character of the gift that is received. For example, two people may be given the gift of teaching, but one may be a more gifted teacher than the other. One teacher may be better at teaching children, whereas their friend is better at teaching adults. The difference in gifts is based on the will of Christ. We are to have unity, but we are not to be identical copies of one another. Notice that verse 7 says that Christ gives

The Library of Celsus in the ancient Greek city of Ephesus (in modern day Turkey). The church in Ephesus was established by Paul on his second missionary journey (Acts 18:19).

grace to people, while verse 11 says that Christ gives people as gifts to people. The gifted people that Christ gives to the church are to help build the church up. We are all to grow together. It is absolutely necessary for us all to cooperate together and use our gifts to help each other grow. The goal is maturity. The goal is experiencing all that there is to experience in Christ. We cannot do this on our own. We need each other. We can only flourish if we grow together. If we truly love each other, we will dedicate our lives to cultivating this kind of pure and holy growth. This is why Paul says that we are to "speak the truth in love" (v. 15), and that the body will grow as we work "in love" (v. 16). Love is essential for spiritual and ethical growth. The quality of a church's love is one way that we can judge its spiritual health.

## EPHESIANS 4:17-24

The way of living that Paul describes in 4:1–16 (which is how we begin to live a life worthy of the calling that we have received), is completely antithetical to the way we were living before we were saved. If we want to understand how strong the contrast is, we need to read Ephesians 2:1–10. Paul knows that if we are going to live the way that God wants us to, we are going to have to cultivate certain virtues and get rid of certain vices (cf. Colossians 3:5–17). There are many other passages we could consider, but at this time we will be constrained to what Paul writes in Ephesians 4:17–24:

> So I tell you this, and insist on it in the Lord, that you must no longer live as the Gentiles do, in the futility of their thinking. They are darkened in their understanding and separated from the life of God because of the ignorance that is in them due to the hardening of their hearts. Having lost all sensitivity, they have given themselves over to sensuality so as to indulge in every kind of impurity, and they are full of greed.
> 
> That, however, is not the way of life you learned when you heard about Christ and were taught in him in accordance with the

truth that is in Jesus. You were taught, with regard to your former way of life, to put off your old self, which is being corrupted by its deceitful desires; to be made new in the attitude of your minds; and to put on the new self, created to be like God in true righteousness and holiness.

Living a holy and ethical life that pleases God requires us to get rid of old ways of living. Paul tells his readers that they must not live as they used to do before they were saved. He makes a connection between their old pattern of life and their old way of thinking (which he describes as futile). Their worldview was so distorted that they acted like fools. Notice the terribly sad description of their state of existence:

1. Their thinking was futile (v. 17).
2. Their understanding was darkened. They had no light in their minds. As a result, they could not comprehend God's truth and the reality in which they were living (v. 18).
3. They were separated from the life of God. Their life, just as it is described in Ephesians 2:1–2, was a living death. The only life that is really life is that which is found in God. But because of their sin, they were separated from it.
4. They were ignorant because of their hardened hearts (v. 18). They stubbornly refused to acknowledge the truth and light of God that surrounded them. Just like it says in Romans 1:18, they suppressed the truth in their unrighteousness. This ignorance was not a matter of being innocently mistaken. Rather, it was a matter of intentionally hardening their hearts. They did not see because they did not *want* to see. Since they were entirely responsible for their state, they were without excuse.

   > **They did not see because they did not want to see.**

5. They lost all sensitivity and increased their guilt (v. 19). When Paul says that they "lost all sensitivity" he uses a term that denotes tough callouses on the skin. Over time, a callous can form on our hands so that they no longer blister when we work hard. The callous renders our skin insensitive. Unfortunately,

we can also build up callouses on our hearts and consciences so that we become insensitive to sin. Once the heart is calloused, people can give themselves over to more and more sin and folly because they do not feel the shame and remorse they ought to feel for their evil conduct. Instead of practicing the ethics of virtue, they practice the immorality of depravity. Every kind of impurity becomes an option for them, and they are greedy for ever-increasing measures of it.

It is no surprise that in verse 17 Paul insists that believers need to stop living that way! He goes on to provide a beautiful image of the believers clothing themselves with Christ and virtue as they throw away their old and defiled garments. Those who are lost live in a delusion; in Christ, sanity is restored because we discover the *truth* that is in Jesus. Now that we can see, we have a responsibility to put off the old self and all of its ways. Deceitful desires—that is, desires that lie to us with their false promises—need to be jettisoned forever and replaced with goodness and truth. We are called to renew our minds, learning to think God's thoughts after him. There are new principles and standards for us to learn and to practice. Just as we are called to *put off* the old self (v. 22), so we are called to *put on* the new self (v. 24). This new self is not merely a slight improvement over the old! The old self was dead in trespasses and sins (Ephesians 2:1)—it was blind, ignorant, calloused and brutal. The new self is created to be like God. This does not mean, of course, that the new self is omnipotent or eternal. What it means is that the new self is like God in terms of having true righteousness and holiness. God is truth, and it is his truth that makes us holy and righteous. Our new self is set apart and reserved for God, and we are now on the right side of God's law. We are to live out this new reality and put it into practice. We can only live a life worthy of the calling we have received by living in love, truth, holiness and righteousness. In the next passage, Paul turns his attention to describing in more particular detail what such a life will be like.

## EPHESIANS 4:25-32

Therefore each of you must put off falsehood and speak truthfully to your neighbor, for we are all members of one body. "In your anger do not sin": Do not let the sun go down while you are still angry, and do not give the devil a foothold. Anyone who has been stealing must steal no longer, but must work, doing something useful with their own hands, that they may have something to share with those in need.

Do not let any unwholesome talk come out of your mouths, but only what is helpful for building others up according to their needs, that it may benefit those who listen. And do not grieve the Holy Spirit of God, with whom you were sealed for the day of redemption. Get rid of all bitterness, rage and anger, brawling and slander, along with every form of malice. Be kind and compassionate to one another, forgiving each other, just as in Christ God forgave you.

Having established the general principle that we are to put off our old way of living and put on the new self that is created to be like God in true righteousness and holiness, Paul introduces an extended discourse on specific ethical matters. The opening word "therefore" signals that the following material is based on what the apostle has just finished saying. Everything that he is about to say is grounded in this great life change which is ours by the grace of God in Christ Jesus our Lord.

Since we are not to live as we used to, we must stop lying to one another and start speaking truthfully. Notice that this command is phrased in both a negative and a positive way. (Negative: stop lying! Positive: speak the truth!) One of the reasons why this is so important is that we are all members of Christ's body. Imagine what would happen if some of your body parts started lying to each other. What would happen if your eyes saw that a bottle was filled with poison, but then they lied to your hand so that you picked it up and drank it? Obviously this would affect the lying eye as well as the rest of the body! There is no way that a body could survive if its members were acting to deceive each other. By analogy, the new covenant community—the body of Christ—cannot survive and flourish in an atmosphere of lies and deceit.

**The body of Christ cannot survive and flourish in an atmosphere of lies and deceit.**

Telling the truth might seem like a very basic ethical duty, but in many cultures lying and deceit are incredibly prevalent, and some forms of lying are even widely accepted. In the Western World, people lie to each other all the time. This does not always take the form of a person saying, "Y didn't happen" when that person knows that Y did in fact occur. What is far more common is people exaggerating the truth to make it something that it really isn't. Or they may minimize certain facts in order to present themselves in the best possible light. Managing a public image is something that many people try to do, and the tactics they use are often less than fully honest. We are to learn to speak the truth to one another. The truth is not to be shaded to our own advantage. But we are also to remember Paul's words in 4:15 that we are to be, "speaking the truth in love." People can say things that are true, but they can say them to be cruel or hurtful. Sometimes the truth can be spoken at the wrong time and in the wrong way. Sadly, the truth can be a weapon that hurts people deeply. Believers are not to lie, but they are also not to use the truth to harm others. We need to speak truthfully to one another, but we always need to speak the truth in love. The ethics of truth-telling involves putting away all forms of deception, speaking the truth and anchoring it all in the loving unity of the body of Christ.

Since we are all members of one body, and we are charged with keeping the unity of the Spirit in the bond of peace, we must control our tempers and anger. Paul tells us in verse 26 that we must not sin in our anger (quoting Psalm 4:4). There are times when God is angry, and there were times in Jesus' life when he grew angry as well. Being angry, then, does not necessarily mean that you are sinning. In fact, given the brokenness and pain in our world, there are times when it is entirely right to be righteously angry. We need to be honest, however; the truth is that we are almost never *righteously* angry. Even when a response of righteous anger is called for, we are so self-centred and sinful that we will almost certainly—at best—experience a mixture of righteous and unrighteous anger. In our daily lives, however, virtually all of our displays of anger will be the result of peevishness, irritability, selfishness, ego, and lack of both patience and self-control. James insightfully points out that, "human anger does not produce the righteousness that God desires" (James 1:20).

To sink this point in deeper, Paul repeats a proverbial expression that tells us not to let the sun go down on our anger. This is not to be taken with an extreme literalness. The verse can hardly be taken to mean that if you start fighting early in the morning you get to be angry all day, but if you start fighting right before darkness falls you can only be angry for a short period of time! (Furthermore, what about the countries where they have short days in the winter and long days in the summer? Or what about the places that practice daylight saving time: Do those people get an extra hour to be angry depending on the season of the year?) All such interpretations are silly. The main point of what Paul is saying is that you need to deal with your anger properly and speedily. There is no place for holding a grudge or nursing anger for a long period of time. On the other hand, even if we calm down quickly after an outburst of anger, we have sinned if our anger was not righteous and justified. Self-control and patience are essential for a godly life, and we are to ensure that we do not sin in our anger.

Verse 27 warns the reader to avoid giving the devil a foothold in their life. This remark is not random or foreign to the context. Giving Satan a foothold in your life is directly connected to unresolved anger and outbursts of temper. A foothold is a place to stand, or a place where one has leverage. When Satan has a foothold, he has a home base from which he can launch attacks against us. This allows the devil to have influence in our lives. When we sin in unrighteous anger, we not only harm those around us, we harm ourselves, too. A dangerous and downward spiral is established: the more we sin the more Satan can influence us, and the more he influences us the more we may give in to temptation. As a result, our moral character erodes further and further. It is sobering to realize that the devil can gain a foothold in our lives in many different ways besides uncontrolled anger, but anger is the specific problem that is identified in this text.

> **When Satan has a foothold, he has a home base from which he can launch attacks against us.**

Moving away from anger and the devil, Paul addresses thieves in the church, and tells them that they must stop stealing! If there are some believers in the fellowship who used to steal, they have an ethical obligation to do so no longer. This might seem like it should go without saying, but the Christian community is sometimes very far from perfect maturity, and there will always be new converts who are just learning how to put off the old ways and put on the new self. Fallen societies provide very different moral communities than what we ought to find in the church. "You shall not steal" is one of the Ten Commandments, and it is a basic principle of God's old covenant law and Christ's standards for his new covenant people.

Notice, however, that simply refraining from stealing is not sufficient for Christian morality. Not only is stealing forbidden, but the one who used to steal is now told to get a job and earn an honest living. They must do something useful; they must contribute and be productive. Formerly, the thief stole from the person who had worked hard and had earned their possessions. Paul commands the thief to stop stealing and to earn an honest living, but then he goes one step farther. Paul goes on to add that one of the great motives for having a job is so that the worker can share with those who are in need. This is a complete reversal for the thief; it is a total life transformation! The thief used to *take* from the person who had earned their goods honestly, but now the thief earns their own goods honestly so that they can *give* to those who need help. The taker has now become a giver: the thief who took things dishonestly now chooses to give away what they have rightfully earned. These principles do not give us a whole Christian ethic for work, but they do contain an important principle. Neither selfish capitalist materialism nor selfish socialist materialism is the motive behind Christian work that pleases God. We do not work merely for ourselves—we work for both God and others. (We elaborate on these themes in chapter 8.)

Christian morality and ethics include both what we do and also what we say. In verse 29, Paul instructs believers not to allow any unwholesome talk to come out of their mouths. The language he uses is very strong: not one single word that ever comes out of our lips is to be harmful. It is better to say nothing than to speak words that are spiritually harmful. Each and every word the believer utters is to be edifying and wholesome. Our words are to impart life and health to everyone who listens. It is very important that the words we use are profitable for building up those who hear us. The only way our words will build others up is if we understand what people need. Pastors and church leaders need to learn how to feed their people. Infants need milk, not meat.

> **The only way our words will build others up is if we understand what people need.**

There is a time and a place for rebuke, and there is a time and a place for correction, but there is also a time and place for praise and encouragement. Assessing the needs of everyone we are talking to, we should choose each and every word with the intention of building them up. Since an unwholesome word can never accomplish this positive goal, not a single unwholesome word should ever come out of our mouths. Let us speak the truth in love at this point: If this one verse was applied rigorously, an unimaginably large percentage of things that are voiced every day would never be said. Our society, churches and families would be instantly transformed if unwholesome words were never spoken. There would also be an incredibly refreshing increase in golden and comfortable silence.

Connected to unwholesome talk is the possibility of grieving the Holy Spirit of God (v. 30). When we speak in ways that are impure and unholy, God's Holy Spirit cannot be comfortable or pleased. When our speech tears others down and hurts the community, God's Spirit is affected. He is a person and experiences personal emotions and relational dynamics. When we speak in an unwholesome manner we do not just hurt our human listeners, we grieve God himself. If we love God, this is a great motivator to fulfil God's commands for holy and ethical speech. We guard our tongues because of our love and reverence for the Spirit.

> **When we speak in an unwholesome manner we do not just hurt our human listeners, we grieve God himself.**

Paul continues in 4:31 to list other vices that we need to eliminate from our lives. The list includes "all bitterness, rage and anger, brawling and slander, along with every form of malice." Many of these specific items have already been eliminated logically by the previous verses. If we have already been told not to sin in our anger, then rage, unrighteous anger and brawling are clearly ruled out. Since slander cannot possibly be considered wholesome or edifying, it too has already been eliminated. Malice is born from an evil and ill will, and thus must not be found in the believer. In contrast to these vices and evils, Christians are to be kind, compassionate and forgiving (v. 32). Jesus Christ himself sets the standard for forgiveness. No matter what anyone has ever done to us—and we recognize that people can do unspeakably terrible and evil things to one another—it is a fact that nobody can sin against us to the extent that we have sinned against an infinitely high and holy God. And the sins of others can never cause us to suffer more than Christ suffered on the cross when he paid the penalty for our sins. When we are offended—when we are sinned against—we are to remember our Lord, the salvation blessings listed in Ephesians 1:3-14, the extent of his sovereign love, mercy, grace, kindness and compassion, and we are to forgive as the Lord forgave us. If we did this, bitterness, animosity and estrangement would die forever, and the new covenant community would be a paradise.

## EPHESIANS 5:1-20

Continuing with the theme of looking to Christ as our standard for ethical conduct, Paul writes:

> Follow God's example, therefore, as dearly loved children and walk in the way of love, just as Christ loved us and gave himself up for us as a fragrant offering and sacrifice to God.
>   But among you there must not be even a hint of sexual immorality, or of any kind of impurity, or of greed, because these are improper for God's holy people. Nor should there be obscenity, foolish talk or coarse joking, which are out of place, but rather thanksgiving. For of this you can be sure: No immoral, impure or greedy person—such a person is an idolater—has any inheritance in the kingdom of Christ and of God. Let no one deceive you with empty words, for because of such things God's wrath comes on those who are disobedient. Therefore do not be partners with them.
>   For you were once darkness, but now you are light in the Lord. Live as children of light (for the fruit of the light consists in all goodness, righteousness and truth) and find out what pleases the Lord. Have nothing to do with the fruitless deeds of darkness, but rather expose them. It is shameful even to mention what the disobedient do in secret. But everything exposed by the light becomes visible—and everything that is illuminated becomes a light. This is why it is said:
>
>> "Wake up, sleeper,
>>   rise from the dead,

and Christ will shine on you."

Be very careful, then, how you live—not as unwise but as wise, making the most of every opportunity, because the days are evil. Therefore do not be foolish, but understand what the Lord's will is. Do not get drunk on wine, which leads to debauchery. Instead, be filled with the Spirit, speaking to one another with psalms, hymns, and songs from the Spirit. Sing and make music from your heart to the Lord, always giving thanks to God the Father for everything, in the name of our Lord Jesus Christ.

The first two verses of Ephesians 5 are infused with love and a call to imitate the God who is love and the Christ who displays love perfectly. We cannot follow God's example in everything since he is our infinite Creator, but we are called to follow his example in love. If we are to live a life worthy of the calling that we have received, we must walk the path of love. When we do so, we find ourselves following in the footsteps of our Lord and Saviour. Because God loved us he sent his Son into the world (John 3:16), and because Christ loved us he willingly came to earth to give himself up for us. The path of love that Jesus walked was a path of holy sacrifice. When Paul says that Jesus gave himself "as a fragrant offering and sacrifice" (5:2), he is using language that summarizes the acceptable sacrifices of the Old Testament sacrificial system. This was done *for us*. When we recall the blessings of Ephesians 1:3–14, and we connect them to this text, we are being told that every blessing that is ours in Christ has come through his self-sacrifice, and his self-sacrifice came about because of his love. That love is our standard; that love is our path. We cannot live a holy and ethical life unless we imitate God, walk in the way of love and love as Christ loved us.

**The path of love that Jesus walked was a path of holy sacrifice.**

From the beauty, purity, goodness and love of Christ, Paul moves to the defilement and selfishness that God's holy people must avoid. Among God's holy people (i.e., people who are reserved only for God and who live in the realm of the sacred), there is not to be even a suggestion of sexual immorality. Given the surrounding pagan culture, it is difficult to exaggerate the significance of this statement. We have, however, discussed sexual ethics in Chapter 5, so we will hasten on. The only thing that needs to be said at this point is that the noble and self-giving love of Christ stands in utter contrast to the selfishness that lies at the heart of all sexually immoral behaviour. "Impurity" includes sexual immorality, but it is probably a more general reference to anything that defiles or is unclean. "Greed" is also general, but it likely includes the type of insatiable desire that feeds sexual lust. Whereas Christ in holy love gave himself over to death for us, sexual immorality, impurity and greed are entirely self-centred, unholy and unloving. They take advantage and abuse, rather than give and bless. As much as the pagan world may revolve around such things, they are improper for God's holy people.

> ... the noble and self-giving love of Christ stands in utter contrast to the selfishness that lies at the heart of all sexually immoral behaviour.

Vulgar speech is also out of place for God's holy people (5:4). The type of speech mentioned in this verse is, of course, already ruled out by 4:29. But given the context, Paul is elaborating on the impurity and defilement that we need to get rid of. Obscene speech is the kind of speech that flows out of an unclean heart and mind. It is vulgar and irreverent. It takes something which ought to be treated with holiness—like sex, for example—and debases it and makes it something dirty. It is lewd and uncouth. "Foolish talk" is not a category that forbids making jokes or having a sense of humour. It does not mean that you must always be serious or dour. It does mean, however, that you must avoid insipid talk that is negative and unedifying. There are innumerable conversations that go on in our world that are simply not worth joining. They waste time and their content is harmful. "Coarse joking" is rude humour. It may be obscene and sexual, or it may be cutting and genuinely hurt others. Sometimes a joke can be told that hurts someone deeply, even though others laugh. Jokes that ridicule can ruin relationships, and one person may weep while the crowd laughs. In the new covenant community, there is no room for such things; they are out of place. What does belong is thanksgiving. We are called to replace harmful, unedifying, hurtful, rude and obscene words with words of thankfulness. The ultimate antidote to unwholesome speech is not silence—it is praise and thanksgiving, which are words of love.

Just in case we think that such vices are trivial, in 5:5–7 Paul explains how serious they are. He categorically states that the immoral person will not have a place in the kingdom of God. Those whose lives are characterized by ethical impurity prove that they do not belong to Christ, and as a result they are excluded from the blessings that are found only in him. The greedy individual is an idolater, since they want the entire world to be organized around their illicit desires. Rather than receiving blessings from the hand of God, they resent the fact that they don't have everything they want. Their desire is to be sovereign, usurping the place of God. This is idolatry, since it puts something else (i.e., themselves) on God's throne.

Paul acknowledges that many people will deny that these attitudes and practices will result in judgement, but he warns his readers not to be deceived. God's wrath will in fact come on those who disobey his law and Word. Denial does not alter reality. Judgement is so certain that Paul warns us not even to partner with those who are disobedient (i.e., we are not to share in their sins). Whether someone calls themselves a Christian is irrelevant—all those who are living a life that is characterized by unrepentant immorality prove that they do not know God through Jesus Christ. We are not saved *on the basis* of our ethical conduct and moral righteousness, but a holy life is necessary *evidence* of the presence of the Spirit, who is the seal of our redemption. A life marked by the absence of ethical purity is a life that demonstrates the absence of Christ and his saving grace.

> **A life marked by the absence of ethical purity is a life that demonstrates the absence of Christ and his saving grace.**

Verses 8–14 provide more reasons for shunning evil and cultivating the fruits of ethical goodness. Genuine Christians have been called out of darkness (cf. Ephesians 2:1–10) and are now light in the Lord. Actually, what Paul says is even stronger—we were not only in an atmosphere of darkness, we *ourselves* were once darkness. But by grace, we are now the light of the Lord. The fruit of this light is seen in goodness, righteousness and truth. Moral goodness leads to acts of righteousness, which are in line with ultimate reality and truth. This fruit of light is obviously very general, and as such will be manifested in a wide variety of ways. Like the fruit of the Spirit (Galatians 5:22–23), these virtues are fruit which also bear fruit. As fruit contains the seeds of the next fruit crop, so the fruit of light produces a cycle of moral fruitfulness. Once we begin to cultivate an orchard of moral fruit, we will begin to find that the more we harvest the more the crop increases. Paul goes on to say that we are to learn and study what pleases the Lord, and then act in response to him. Our new life of light stands in absolute contrast to our old life of darkness. In our darkness, we did not know how to please the Lord. Now we do. Our lives ought to demonstrate that complete change in orientation.

**We shine the light into the dark corners so that sin is revealed and sinners can see their need of repentance.**

Even though we are light in the Lord, we are imperfect, and we live in an imperfect and fallen world. Given the reality of sin and darkness, how should we act when we are confronted with evil? First, we ought to look critically at our own lives to make sure that we are producing the fruit of light and utterly shunning the fruit of darkness. Second, we ought to expose evil to the light. When light shines into the darkness, it is darkness that is defeated. Light scatters darkness and ends its tyranny. We often find that evil likes to hide its face. As Paul says, the things that evil people do in secret are shameful. It's tragic that we even need to discuss it. But evil is not going to go away when it is allowed to hide in a dark corner; it must be exposed. We shine the light into the dark corners so that sin is revealed and sinners can see their need of repentance. Christ shines on the dead and they come to life; he shines on the sleepers and they awake. When we are confronted by sin, our desire should be to see the light of Christ prevail over the darkness. It should go without saying that we can't live this way if we are the ones who are cultivating the fruitless deeds of darkness.

Paul draws Ephesians 4:1–5:20 to a close by providing another mixture of general and specific instructions for living a life worthy of the calling that we have received. Unsurprisingly, we are to live *carefully*. (Paul literally says that we are to "walk" carefully.) Life is not a game, and given the nature of darkness and light, as well as the eternal consequences that stand before us, we need to walk on God's path with wisdom. The Book of Proverbs shows us that there are really only two paths—the path of folly and the path of wisdom. Those who have been redeemed must walk wisely.

Those who are redeemed must also redeem their time. ("Redeeming" or "buying up" the time is closer to Paul's original wording.) The image here is of going to the market and buying time from a vendor. Since we can only live one moment at a time, and since our time in this world is finite, we must not waste it. Time is more precious than money or any other material possession. Time well spent is one of the marks of a wise and ethically pure life. One compelling reason to buy up our time and make the most of it is that the days are evil. There is a constant pressure to spend our time the way that the world does around us. But we must resist this temptation—we must use our time for the Lord. Rather than being fools, we need to understand the Lord's revealed will. We need to live every moment in Christ's light.

> **Time well spent is one of the marks of a wise and ethically pure life.**

If we are to live wisely, we will need all of our wits and faculties operating at their fullest capacity. We will need to be alert and sober. Since drunkenness destroys self-control and inhibition, and since it impairs our physical and mental capacities, getting drunk is antithetical to living wisely and walking on God's path. It is not hard to recognize that drunkenness does not generate positive moral behaviour! The time is too short, and the world is too dark, for the Christian to sin by getting drunk. Furthermore, drunkenness is a sin that leads to other sins. As Paul says, it leads to debauchery.

> **...getting drunk is antithetical to living wisely and walking on God's path.**

Being filled with wine to the point of intoxication is forbidden, but there is Someone with whom we are to be filled to the point of overflowing. Having been sealed by the Spirit, and having been baptized with the Spirit (which is a once for all event at conversion), we are to be continually filled with the Spirit as we live our lives. We are to willingly and gratefully submit ourselves to the Spirit's influence and control. As much as we must not allow Satan to have even the smallest of footholds in our lives, we must throw open every corner of our hearts to the Spirit. It is only by the Spirit's power that we will produce the fruit of righteousness. Apart from him, we cannot be wise, nor can we live in a way that pleases and honours the Lord. He is the *Holy* Spirit because his nature is holy and because he makes God's people holy. Success in Christian ethics depends on keeping in step with the Spirit. In the same way that excessive amounts of alcohol can overwhelm an individual, believers are to be overwhelmed by the Spirit. But the resulting fruit could not be more different! Drunkenness is marked by loss of self-control, but self-control is part of the fruit of the Spirit.

Drunkenness is also marked by slurred speech and incoherent thoughts. Those who are filled with the Spirit, however, are marked by clarity and insight into reality, which generates songs of praise to God. Paul tells us to make music to God from the depths of our being, as we are prompted and guided by the Spirit. No matter what our circumstances are, all of our speech and conduct should be filled with thanksgiving to God through Jesus Christ. Following God, imitating Christ and being filled with the Spirit is the path of joy. God's ethical and moral commands do not kill joy; they produce it. Only when we walk in the fellowship of Christ can we discover how beautiful and refreshing holiness is. Living ethically is primarily for God's glory, but it is also for our good. By his grace, let us live a life worthy of the calling that we have received.

> **Only when we walk in the fellowship of Christ can we discover how beautiful and refreshing holiness is.**

## REFLECTION QUESTIONS

1. Was there one part of Ephesians 4:1–5:20 that really stood out to you in terms of your own life? What areas are you weak in?

2. What other biblical passages are relevant to Paul's material in that section of Ephesians?

3. Even though we will never be perfect in this world, are we doing an adequate job of living a life worthy of the calling that we have received? How are our churches doing?

4. Is a transformed life a necessary sign of conversion? Or can someone be saved without showing much change in their ethics and lifestyle?

# PART TWO
# Ethical issues

# 4

# *Family ethics*

As we established in the first section, God is our standard for morality. We need to examine every ethical issue—including the ethics of family life—in the light of his character. God is a perfect being. He is not lacking any good quality, and every quality that he has is possessed maximally (i.e., perfectly). The living God exists in splendour, perfection, glory and absolute blessedness. His nature is unchanging, and it cannot be improved. Due to our sin and imperfections, we cannot imagine what it is like to exist in a state of such blessedness, joy, satisfaction and love. As the triune God, God's love is expressed inside of his own special, internal community. The God who is the three-in-one experiences and enjoys eternal and unbroken personal love, harmony and fellowship. God's manner of existence cannot be equalled or surpassed.

## THE REALITY OF SINGLENESS

Given the fact that God is perfect and perfectly satisfied, it is interesting to observe that he does not have a consort or spouse. God, the most perfect being imaginable, is unmarried. The angels God has created to live in his presence are also unmarried. Furthermore, Jesus taught that all of the redeemed in their state of future glorification will likewise neither marry nor be given in marriage. Jesus said, "At the resurrection people will neither marry nor be given in marriage; they will be like the angels in heaven" (Matthew 22:30). Paul taught that in many ways it was better to be single even in this world than to be married (1 Corinthians 7:32–35). Most importantly of all, *Jesus* never married, and he lived the most complete and God-honouring human life in history. Jesus blessed marriage and recognized its place in God's good design, but he also praised singleness and spent his entire life as a single man.

> **Jesus never married, and he lived the most complete and God-honouring human life in history.**

Sadly, despite these biblical and theological realities, many of our churches have prized marriage and families so highly that they have—sometimes unintentionally, and other times out of inexcusable rudeness—devalued, demeaned and diminished the unmarried adults in their fellowship. Adults who are single can sometimes be made to feel like second-class citizens in their own church culture. Or they may be told that they have instrumental value only. (That is, single people can receive the message that they are valuable for what they can *do* rather than valuable for who they *are*. For example, the unmarried may be seen only as useful because they have more time for working hard and serving in the church.) Sadly, some may even be made to feel that their value is based on their potential to someday find a marriage partner. We need to recognize that every person—whether single or married—has intrinsic value simply by being the image-bearer of God. Those whom Christ has redeemed with his blood are the objects of infinite and perfect love. Every Christian has Christ as their bridegroom, and every Christian has been loved by one who literally gave up his life for them (Mark 10:45).

Before we talk about marriage, then, we need to acknowledge that marriage is *not* a state that everyone should be trying to enter. In fact, remaining unmarried for the sake of Christ and his kingdom may be a richer and more fulfilling way of living than being married. The Bible says so! Many people, however, try to find support for the universal necessity of marriage in the creation of Adam and Eve. In Genesis 2:18, God says, "It is not good for the man to be alone." To rectify Adam's loneliness, God created Eve. This text is often taken to prove that people are meant to be married, and that they will be lonely and living in a way that is characterized as "not good" until they are united with a spouse. This interpretation of the text is massively mistaken. Not only does it ignore the fact that the Scriptures hallow singleness for the Christian, it also reads into the text a number of assumptions that are simply not there. Adam was completely and totally alone: He was the only human being in existence. No other person in the history of the world has been in that state. Adam—and only Adam—spent time as the *only* human being in creation. People don't necessarily need a spouse to escape loneliness, but they do need other people. They need a community. They need family. They need friends. They need relationships with others. They need love. But marriage isn't required for any of these things. Marriage isn't necessary in order to escape loneliness. And, as many have discovered, it is entirely possible to be both married and lonely.

It is true, of course, that God created Eve so that Adam would no longer be lonely, and it is also true that God created Eve to be Adam's wife. Marriage is God's creation, and it can be wonderful, holy and fulfilling! But it is possible that we diminish Eve's value when we see her as primarily Adam's wife, rather than as an image-bearer of God (Genesis 1:27) who was also Adam's human companion. Nobody seems to think that Adam should be known primarily as Eve's husband. They were married, but they were companions. They were the smallest human community the world has ever known. Since God has ordained that human beings will procreate and reproduce, and he has further ordained that this activity will take place inside the covenant bonds of marriage, Adam and Eve had to be married or else there would be no human race.

**Marriage took place before the Fall into sin... God ordained Adam and Eve's marital covenant.**

It is important to recognize that marriage took place before the Fall into sin, and that God ordained Adam and Eve's marital covenant. However, there is no reason whatsoever to believe that everyone would be married if sin had not entered the world. Singleness is not a result of the Fall or the curse. Even if there was no sin in the world, the human community would likely be a place with both married persons and unmarried persons who would never marry. In fact, without sin, it may be the case that there would be more unmarried people than there are currently. After all, love and community would be perfect. We would be attuned to godliness and holiness—perhaps making us more likely to live now the way we will live in heaven—and more likely to live in full dedication of our lives to the Lord. Furthermore, nobody in an unfallen world would marry *for the wrong reasons*. Let us be honest: today, many marriages are entered into for purely sinful and selfish reasons. Lust, money, pride, insecurity, fear and the like are driving forces behind the making of many marriages.

We must remain sensitive to the fact that unmarried people experience their singleness in different ways. Some who never marry are very happy not to be married. Others have a deep longing for marriage but, for a variety of reasons, have never met a covenant partner. There are still others who were married but, due to divorce or death, find themselves forced against their will to be single again. If a child is orphaned, it is entirely natural for them to wish that they had parents; they are not sinning by doing so. Likewise, a single person can accept God's ordering of their lives and yet find that they still have a deep desire for marriage. If this is the case, they have no reason to feel ashamed. In a similar way, a childless couple may long for children but be unable to conceive. The desire for children is not evil, nor is it inappropriate. Sorrow and longing are not inherently wrong. There is a problem, however, if the idea of marriage becomes an idol, or if the single person lives their life focused on what they don't have rather than on all the blessings that they do have. It is not right if a desire for marriage leads to grumbling, hatred, envy, bitterness or overwhelming self-pity. But the difficulty of the single state can be acknowledged, and the desire for marriage validated. This is not incompatible with joyfully loving and serving the Lord, and even with submitting our will to his will for our lives. Parents may experience the death of a child, and although they may eventually find peace and submission to the will of God, they may always ache and wish their child were alive.

> **It is not right if a desire for marriage leads to grumbling, hatred, envy, bitterness or overwhelming self-pity.**

Accepting God's providence does not mean we won't feel pain.

The church has a moral and religious duty to create a loving community where *everyone* is equally valued, appreciated and esteemed (cf. John 13:34–35). If the unmarried do not feel entirely accepted and honoured by the church, the solution is not to try to find them a marriage partner—the solution is to improve the quality of the church's love and care and community. Little could be more unbiblical than to act as if marriage is the only place where genuine love can be found and experienced. The unmarried need to be able to both give and receive love, which is the same need experienced by those who are married. Some of the finest and most loving people in the world are unmarried. Love can exist in many forms and flow through different channels. Love manifests itself in a variety of ways. Marriage is one unique channel for love, but so is friendship. So is the love of siblings or cousins. But the greatest quality of love should be found among Christians, even if they are unrelated by genetics and bloodlines. We are one body. God loves every one of his children, and we must love all those our Father loves. Not everyone will marry, but everyone needs love. Not everyone will have a spouse, but everyone needs a friend. Everyone needs God and a loving community. And only the church—because it is animated by the Father's love in Christ—can play that role.

**The church has a moral and religious duty to create a loving community where everyone *is* equally valued, appreciated and esteemed.**

We must also remember that many people who eventually marry spend years as single adults. (In fact, everyone who marries experiences living in an unmarried state for years before entering their marriage.) As was mentioned, there are also other people who marry, but then find themselves single again, either due to divorce or the death of their spouse. Someone may marry, suffer the death of their partner after just a few years, and then remain unmarried for the remainder of their lives. Clearly such a person was not more valuable when they were married than they are after the loss of their spouse. They are just as human, just as precious and just as much in need of love and community.

Marriage is not what confers value on people, nor is it an accurate measure of a person's quality or character. God does not judge people's worth on the basis of their marital status. Neither should the church. We need to love and celebrate the community of Christ's body—which includes *every* member—more than we celebrate marriage. Not one solitary person should ever be part of the church without both knowing and experiencing what it is like to share in the mutual self-giving of holy love.

> **God does not judge people's worth on the basis of their marital status. Neither should the church.**

It is this type of perfect community love that the angels experience. As Jesus said, the angels are neither married nor given in marriage (Luke 20:35–36), yet doubtless they love. How could they be in the presence of the one who is holy love without being both holy and loving? But Jesus said this about the angels to help us understand what *our* future in glory will be like. In the new heavens and new earth, we will not be losing anything or missing anything. Sometimes believers who are married grieve the thought of no longer being married in heaven. They mistakenly think that they will be *losing* something; the truth is that they will be *gaining* something. First, their relationship with their spouse will be far better in eternity than it could ever be on earth, even though they won't be married. Second, far from losing intimacy with one person, we will be gaining intimacy with *everybody*! In glory, we will discover that we are closer to *everyone* there than we are to *anyone* here. Marriage by definition is an exclusive and limiting relationship. Today, marriage brings two people together, and as a consequence, it also keeps other people apart. In heaven, our relationships will be inclusive and unlimited. Mutual love and tender closeness will only increase our joy; there will be no jealousy. When we stand in glory, we will all be united without any division. Our bonds there will be stronger than those of the very best marriage here. We will all be one, and in our oneness we will be married to Christ.

> **Our bonds [in heaven] will be stronger than those of the very best marriage here.**

Our mode of existence in glory will be better in every conceivable way than it is now. There is no marriage in heaven besides the marriage of Christ and the church. Yet, it is only there that we will experience pure and perfect love! We will revel in the infinite love of God, but we will also love each other. The new covenant community, the body and bride of Christ, will experience *perfect love in community*, and we will not be coupled off. This means that the church at its most blessed is a community of love without the institution of marriage. If that is the case, then surely we need to see that the church ought to be a wonderful and loving community *now*, where every member is loved, and loved *and loved*, regardless of whether they are married or not. Love must be universal in the Christian church. We are not all married, but we are all family. Nobody in the family should feel relationally isolated, unappreciated or unloved. We are brothers and sisters: let us love one another.

## THE REALITY OF MARRIAGE

Although it is a mistake to deify or glorify marriage, it is designed by God and therefore can be holy, good and beautiful. There is nothing wrong with being unmarried, but there is also nothing wrong with entering into marriage. God, Jesus, the angels and the saints in glory are unmarried, but marriage is a state that has been blessed by God in this temporal world. It is for some, but not for all. It is not eternal, but it does have its time and place. As much as we must not diminish those who are unmarried, we must also not diminish those who enter into marriage. Living life as a single person or living life as a married person are both permissible in God's sight. Marriage must not be exalted beyond its due place, but we do need to recognize the goodness of what it is. In its proper place and viewed in a balanced way, marriage is a precious ordinance that is a gracious and special gift from God.

**Marriage is a creation ordinance that is given to the human race, so it is not reserved exclusively for the followers of God.**

Marriage is a creation ordinance that is given to the human race, so it is not reserved exclusively for the followers of God. Christians must only marry other Christians, but non-believers also have the right to marry one another. This does not mean that there are marriages which are not religious—every marriage is religious in one sense, since marriage can only be sealed by the power and permission of God. It is almighty God—and he alone—who created marriage, who upholds his laws concerning marriage and who sanctifies and blesses the marriage union. God is a covenant-making and covenant-keeping God, and as such the human covenant of marriage is something that can only come from him. Every time marriage vows are exchanged and the covenant is formed between husband and wife, God is the witness, he is the authorizing agent, and what he joins together nobody is to separate (Genesis 2:24; Matthew 19:6).

> *It is almighty God who created marriage, who upholds his laws concerning marriage and who sanctifies and blesses the marriage union.*

When I, Steve, officiate wedding ceremonies, I usually recite the following words: "This occasion marks the celebration of love and commitment with which this man and this woman begin their married life together. Marriage is the union of husband and wife in heart, body, soul and mind." *Marriage is the union of husband and wife in heart, body, soul and mind.* It is not merely a relationship of physical union through sexual intercourse. Being "one" is not merely sexual; it is not restricted to becoming one flesh. In marriage, a husband and wife are to enter into holistic unity. This ideal is not always fulfilled, and although there is an objective and given unity in marriage by definition, the subjective amount of unity can vary greatly from one marriage to the next. Some spouses are deeply compatible; others discover they are not.

Throughout history, and across the world today, the reasons why people in various cultures marry are diverse. Some marry for more pragmatic reasons (eg. economic necessity, the need for producing children to work the farm, etc.). Some are married to people they have never met before, since their families select their spouse and arrange the marriage contract on their behalf. Some cultures believe that people should only marry on the basis of feelings, especially the feeling of "being in love." Some individuals seek to marry the person who they believe "completes" them; they marry someone who they think can help them grow or be happy. As already mentioned, many people also marry for more transparently ignoble reasons, like lust or insecurity. Certain cultures have a view of marriage where the institution is designed as a foundational social block for mutual help, comfort and support. Other cultures view it as quite disposable—it is something to be discarded as soon as feelings change, or someone more attractive comes along.

We need to be aware of the presuppositions that our particular culture has concerning the purpose and goals of marriage. It is far more important, however, to be grounded in what Scripture teaches concerning these things. The truth of Scripture transcends cultures and applies everywhere. Whether you live in a rural African village or a highly sophisticated city in Asia, whether your society prizes marriage or disparages it, whether most of your nation's marriages end in divorce or last for life, you need to form your view of marriage on the basis of the Bible's teachings.

### Ephesians 5:22-33

Paul's instructions to husbands and wives, recorded in Ephesians 5:22-33, are crucial for a biblical understanding of marriage. He writes:

> Wives, submit yourselves to your own husbands as you do to the Lord. For the husband is the head of the wife as Christ is the head of the church, his body, of which he is the Savior. Now as the church submits to Christ, so also wives should submit to their husbands in everything.
>
> Husbands, love your wives, just as Christ loved the church and gave himself up for her to make her holy, cleansing her by the washing with water through the word, and to present her to himself as a radiant church, without stain or wrinkle or any other blemish, but holy and blameless. In this same way, husbands ought to love their wives as their own bodies. He who loves his wife loves himself. After all, no one ever hated their own body, but they feed and care for their body, just as Christ does the church—for we are members of his body. "For this reason a man will leave his father and mother and be united to his wife, and the two will become one flesh." This is a profound mystery—but I am talking about Christ and the church. However, each one of you also must love his wife as he loves himself, and the wife must respect her husband.

These may be the most important words in the world when it comes to understanding marriage. Both husband and wife are to focus their marriage on Christ. Submission and headship, love and respect, holiness and sacrifice—Christ is the point of it all. When Paul says that marriage is a profound mystery, he does not mean that marriage is mysterious or hard to understand (although it can be!). When Paul talks about a theological mystery, he is talking about something that could never be figured out apart from the special revelation of its truth by the Holy Spirit. In other words, until Christ came and the Spirit was poured out, nobody could have known that every marriage in human history was supposed to be an enacted prophecy of the special covenant relationship between Christ and the church. Marriage isn't just being used by Paul as a convenient illustration or analogy—God created the very first marriage in such a way that it would serve a prophetic function. Every marriage is to point beyond itself to Christ's covenant union with his bride, the church.

This great principle anchors the marriage ethic. Wives, God has assigned you the role of the church in this prophetic pageant. Does your attitude toward your husband make it look like the church loves and respects Christ? It is important, of course, to remember that this is an analogy: no human husband is Christ, and they shouldn't be treated by anyone—including their wife—as if they were! Wives are not to worship their husbands. Husbands—not actually being Christ—are also not expected to set themselves up as kings over their families. In fact, Paul doesn't tell husbands to *rule over* their wives, he tells them to *die for them*. The incarnate Son of God willingly suffered and died for us—husbands, that's how you are to act toward your wives. Husbands ought to be sacrificing their own wishes and preferences for their wives, and wives ought to be looking to honour their husbands ahead of themselves. If Christian marriages looked anything like this, the world would notice how harmonious marriage could be. Two people, each only looking to put the other first, to love them, to honour them, to keep them, to bless them—what could be better? Marriage is not about the husband and wife—it is about Christ and the church. Since this is the case, marriage cannot

**In fact, Paul doesn't tell husbands to rule over their wives, he tells them to die for them.**

be based on fleeting feelings—it must be built on grace, love and the power of God.

## Divorce

God did not design marriage to end in divorce, and people do not enter marriage expecting to divorce in the future. When human marriages end in divorce, it is never ideal, and in a world without sin divorce would never occur. This does not mean that divorce is never permissible (there are circumstances in which the Bible allows it). Sadly, whenever there is a divorce there is sin—either because the divorce isn't biblically permitted or because of sinful behaviour which constituted legitimate grounds for the divorce. Some people pursue divorce because they are unsatisfied and lonely, or because they find their marriages unbearably painful. Even if biblical warrant does not exist, some spouses are treated terribly by their partner, and they long for release. The church does not need to condone every reason why someone gets divorced, but there needs to be compassion and grace. Divorce is not an unforgiveable sin. In some marriages, spouses sin against each other hundreds of times a day in word and deed. God hates divorce but that does not mean that every marriage is somehow honouring to him. While some people need to repent for choosing divorce without biblical warrant, others need to repent for their ongoing conduct within their marriage. In both cases, we can be thankful God is merciful.

The most important biblical text on divorce is in Matthew 19:3–9, which records an exchange that took place between Jesus and the Pharisees. The text says:

> Some Pharisees came to him to test him. They asked, "Is it lawful for a man to divorce his wife for any and every reason?"
>
> "Haven't you read," he replied, "that at the beginning the Creator 'made them male and female,' and said, 'For this reason a man will leave his father and mother and be united to his wife, and the two will become one flesh'? So they are no longer two, but one flesh. Therefore what God has joined together, let no one separate."
>
> "Why then," they asked, "did Moses command that a man give his wife a certificate of divorce and send her away?"
>
> Jesus replied, "Moses permitted you to divorce your wives

because your hearts were hard. But it was not this way from the beginning. I tell you that anyone who divorces his wife, except for sexual immorality, and marries another woman commits adultery.

It is important to notice that when Jesus is asked about reasons that would allow for divorce, he begins by reminding his hearers about what marriage is supposed to be. You cannot understand when a marriage can end unless you understand what marriage is in the first place. This also subtly shifts the ground. Instead of worrying about divorce, we should concern ourselves with how to keep marriages intact and how we can uphold God's original design (v. 4–6).

Jesus reminds us that the first thing to remember about marriage is that it is not merely a human institution, nor is it a contract that society invented. Marriage was God's idea (and God's ideas are always good). God has made the married couple one, and what God has united, no one is to separate (v. 6). This speaks of the permanency and importance of the marriage union. Jesus acknowledged that Moses permitted divorce because of the terrible hardness of the people's hearts, but divorce was never the ideal. Far from divorce being acceptable for any reason, Jesus says that it is permissible only in the case of sexual immorality (v. 9). A married person who commits sexual immorality has so violated their marriage vows and covenant that the offended party has the right to divorce. The innocent person isn't breaking the covenant—it is already broken by the one who committed sexual immorality.

In 1 Corinthians 7:15, Paul adds another circumstance in which divorce can occur. Paul says that a believer is not bound in marriage if their spouse abandons them. There are times when one person is wanting and willing to continue to be married, but the other spouse leaves or forces a divorce. There is nothing that can be done in such a circumstance, and the innocent party is not responsible for the divorce. As is the case whenever there are biblical grounds for divorce, the innocent party may remarry provided they do so in the Lord.

Although neither Jesus nor Paul reference spousal abuse, church leaders may need to be involved in cases where one spouse's behaviour places their partner in danger. If a husband is physically abusing and hurting his wife, he is violating his vows and functioning as the opposite of what a husband should be. If we know that a husband is physically hurting his wife, we ought to contact the police and all other relevant authorities or agencies that can help. We should do what we can to help the wife find somewhere to live where she can be safe. Church leaders must do all that they can to protect the victims of spousal abuse. They should also provide biblically-based counselling to help bring healing.

This brings up an important issue. The church, as a loving community, has a great responsibility to protect those who are vulnerable. In the case of spousal abuse, the church needs to intervene (and bring in lawful civil authorities as well). Christians must stand up for the rights and safety of women. The church must not be a place where men feel free to abuse their wives, or to demean any woman in any manner. Spousal abuse is evil, and it completely misrepresents the love of Christ for his church. It must not be tolerated in any form or for any length of time. If there is one place on earth where women should be able to feel loved, valued, cherished and protected, it should be the church. Brothers: Help keep your sisters safe.

We need to remember that, even in circumstances where divorce is *permissible*, it does not follow that it is *necessary*. A husband or wife may fall into adultery, and then the offended partner has every right to proceed with divorce if they so choose. But the offended partner also has the right to forgive and to work to rebuild the marriage relationship. We must avoid putting all of the responsibility for this work on the innocent party, however. Too often one spouse will fall into sexual immorality, but then it's the innocent spouse who is pressured by the church to make the marriage work (and often made to feel guilty or ashamed if they can't). They can sometimes receive the message, "Yes, your partner was unfaithful, but if *you* can forgive them and if *you* want it to work, there's no reason for the marriage to end in divorce." There is a sense in which this is true, of course, but there is another sense in which it is completely wrong. If the marriage ends in divorce, it is the fault of the unfaithful partner who broke their covenant vows, not the faithful partner who upheld their promises! Jesus allows for divorce in certain circumstances—we must not think that we are more righteous than Jesus. The church needs to provide love, support and counsel to married couples, but part of this support may be assuring a believer that Jesus has given them the right to divorce in certain warranted circumstances. And if they listen to Jesus, they are not sinning.

### Children

The Bible has a lot to say about the value and blessing of children:

> God blessed them and said to them, "Be fruitful and increase in number" (Genesis 1:28).

> Children are a heritage from the LORD,
>     offspring a reward from him.
> Like arrows in the hands of a warrior
>     are children born in one's youth.
> Blessed is the man
>     whose quiver is full of them.
> They will not be put to shame
>     when they contend with their opponents in court.
> (Psalm 127:3–5)

> "Whoever welcomes one of these little children in my name welcomes me; and whoever welcomes me does not welcome me but the one who sent me" (Mark 9:37).

> When Jesus saw this, he was indignant. He said to them, "Let the little children come to me, and do not hinder them, for the kingdom of God belongs to such as these" (Mark 10:14).

Just like adults, children bear the image of God. As a result, a child has the same intrinsic value as an adult. Children are not merely valuable because of their potential for future work, productivity or service. They are valuable simply because they are human, and human beings are valuable simply because of their God-given nature. Since the Bible highly values people, it highly values children.

In the biblical culture—as in many economies today—one of the reasons why children were considered assets was because of the contributions they could make in the household labour force. Children were expected to work productively and helpfully, and as they grew older, stronger and wiser, they would take on more and more responsibility. As parents aged and grew infirm, their older children would end up caring for them. So children were also seen as a security for old age. Families, not government programs, were expected to care for the elderly. Parents would care for their young children, and then the time would come when grown children would care for their aged parents. Nevertheless, children are not merely valuable because of what they can *do* either now or in the future—they are valuable because of what they *are* by virtue of God's creation.

The first command and blessing given to the human race was to be fruitful and multiply (Genesis 1:28). It is important to recognize that this command was given to the race as a whole, and not to every individual. We have already seen that not everyone will—or should—marry. Since God's design is for children to be born to married couples, this will mean that not every individual is expected to produce children. God calls everyone to flourish in fruitfulness, love and blessing, and neither marriage nor offspring are required to fulfil this calling. God's ideal for children, however, is that they be born into a covenant context. He wants children to be born to a man and a woman who are committed and bound in covenant to one another. The solidity and love of the parents provides a foundation for the security and nurturing of the

***God's ideal for children…is that they be born into a covenant context.***

child. In a situation where one person ends up with the sole care of the child (eg. because of death or abandonment), God can give the grace and strength that is needed. Children can still flourish when raised by a loving single parent. The Christian community should also be quick and active in helping to support single parents.

Parents have an absolute ethical responsibility to care for their children. Paul declared, "Anyone who does not provide for their relatives, and especially for their own household, has denied the faith and is worse than an unbeliever" (1 Timothy 5:8). Neglecting a child's needs is inexcusably wicked. A parent's responsibility does not end with the bare provision of physical needs like food, clothing and shelter. Parents are responsible before God for raising and nurturing their children in moral goodness, and directing them to honour the Lord. Texts could be multiplied, but these three will be sufficient:

> Hear, O Israel: The Lord our God, the Lord is one. Love the Lord your God with all your heart and with all your soul and with all your strength. These commandments that I give you today are to be on your hearts. *Impress them on your children.* Talk about them when you sit at home and when you walk along the road, when you lie down and when you get up (Deuteronomy 6:4–7, emphasis added).

> Start children off on the way they should go,
>    and even when they are old they will not turn from it.
> (Proverbs 22:6)

> Children, obey your parents in the Lord, for this is right. "Honor your father and mother"—which is the first commandment with a promise— "so that it may go well with you and that you may enjoy long life on the earth."
> Fathers, do not exasperate your children; instead, bring them up in the training and instruction of the Lord (Ephesians 6:1–4).

Notice it is the parents' responsibility to teach their children about the Lord, and to start them off on the right path. The passage in Deuteronomy envisions a family dynamic where the things of God are discussed throughout the day as occasion arises. All of life is to be lived in religious consciousness, and our thoughts and actions are to be directed toward honouring God in all that we do. Paul's words in Ephesians are very interesting, since he addresses children directly and tells them that they have responsibilities to their parents. Children are indeed responsible moral agents who are expected to obey God's Word and to be obedient to their parents. Parents, however, also have responsibilities to their children. Fathers are told not to exasperate their children, but rather to teach them and train them in the things of the Lord. It is instructive that the one thing Paul told fathers to refrain from was "exasperating" their children. When parents are overly critical, set impossibly high standards, are verbally or physically abusive, operate inconsistently or fail to love their children, the children suffer.

Children also suffer, however, from a lack of boundaries and discipline. Discipline involves loving, wise and firm guidance and correction (Hebrews 12:5–6). Parents are not to be harsh or cruel, but neither are they to be apathetic or overly indulgent. Children are not meant to raise themselves, and they are certainly not supposed to domineer in the household. God has given parents authority over their children, and this authority is to be used in conjunction with love and care. Failure to properly and lovingly discipline is a failure of love. In fact, the Bible teaches that where discipline is absent, love is absent. As Scripture says,

> **Discipline involves loving, wise and firm guidance and correction.**

> Endure hardship as discipline; God is treating you as his children. For what children are not disciplined by their father? If you are not disciplined—and everyone undergoes discipline—then you are not legitimate, not true sons and daughters at all. Moreover, we have all had human fathers who disciplined us and we respected them for it. How much more should we submit to the Father of spirits and live! They disciplined us for a little while as they thought best; but God disciplines us for our good, in order that we may share in his holiness. No discipline seems pleasant at the time, but painful. Later on, however, it produces a harvest of righteousness and peace for those who have been trained by it (Hebrews 12:7–11).

Although the main point of this passage is that God the Father disciplines his children, the analogy is drawn with human parenting. If there is no discipline where it is needed, it is a sign that something is fundamentally wrong in the relationship. We can tell we are God's children because he loves us and disciplines us. Human parents should learn to imitate God the Father by loving and disciplining their children. Discipline always seems unpleasant, but the short-term pain is worth the long-term good that it can produce. It is vital that parents avoid the common errors of either being too harsh or too indulgent.

### Child abuse

Children are vulnerable. When they are born, children are completely helpless and dependent. They will die unless they are given constant care. It takes years and years of growth before children can support themselves. Parents have far more physical, mental, economic and social power than their children. This means that parents are in a position where they can either do their children tremendous good or tremendous harm. One of the saddest tragedies of all is that some parents abuse their children. The number of children who suffer abuse at the hands of their own parents is unimaginable. It is a global nightmare, and Christians have an ethical responsibility to help the helpless and combat it.

Child abuse can involve excessive or cruel physical, psychological or emotional punishment. It can also involve neglecting children's physical needs (and sometimes emotional needs as well). One particularly horrific form of child abuse is sexual molestation. We discuss sexual ethics in another chapter, but since any sexual activity between unmarried people is sinful, it is obvious that sexually abusing a child is forbidden by God. But sexually abusing a child is even worse than engaging in illegitimate consensual sexual activity with another adult. Child sexual abuse is exploitation of the vilest kind. In our local church, we have made it clear that if someone sexually abuses a child—whether their own child or someone else's—we will not only discipline them as a church, we will call the police and do whatever we can to ensure that they are punished by the civil authorities and the judicial system. We will also do whatever we can to provide care and counselling for the victim. There must be no tolerance whatsoever for child abuse in our churches.

> **We must also do whatever we can to stop child abuse in our wider cultures and societies.**

We must also do whatever we can to stop child abuse in our wider cultures and societies. There is one practice in particular which has been receiving media attention and which the church must oppose. This practice is *female genital mutilation*. It is sometimes referred to as female circumcision, but while a case can be made that male circumcision provides a health benefit, there is no health benefit for females—in fact, it is entirely harmful, destructive and degrading.

This abusive practice involves the removal of a young woman's (or young girl's) external genitalia. There is literally *nothing* positive that results from this practice, and the negative effects are enormous. It is horrendously painful and can produce extreme complications (including infection that leads to death). The physical and psychological scars may never fade. Whether the rationale for the practice is cultural or religious, it is barbaric. It is certainly against all good medical wisdom. More importantly, it is also diametrically opposed to the principles of God's Holy Word. God has designed women with proper female genitalia, and it is perversely evil to mutilate them. If a church is in an area where this grotesque practice is known to be happening, part of the church's ethical responsibility there is to actively oppose it and to provide protection for those seeking to escape from it. Even if one girl in the world experienced this abusive procedure, it would be one girl too many. Given the number of girls who are victimized by this practice, this is an ethical issue that Christians must not ignore.

### *Fertility and technology*

Is it ethically acceptable for a couple to use medical technology to aid in the conception of a child, if they have not been able to conceive through normal sexual intercourse? On the other side of the equation, is it ethically acceptable for a couple to use medical technology to *avoid* conceiving a child? The practicality of such questions will vary enormously depending on where you live. For some people, either the medical resources for such procedures are non-existent, or they are simply unaffordable even if available. Other people, however, will have both the access to the necessary medical technology, and the financial means to pay for it. But just because we *can* do something doesn't make it right. So how should Christians think about the relationship between reproduction and technology, especially given the fact that the Bible does not address modern technology?

> *...just because we can do something doesn't make it right.*

In some ways, the question can only be answered inside of a wider understanding of what the proper roles for medicine, science and technology is. This is a fallen world, and medical advancements are one way that we can and should help bring healing and restore wholeness. If a person is blind, but a surgical procedure could cause them to see, there would be nothing wrong with having the surgery. Likewise, if a person has a disease that medication can cure, it is not immoral to take the medicine. This simple line of reasoning extends to issues of biological reproduction. If a couple is not able to conceive, there is nothing wrong *in theory* with using medical procedures to help bring about a conception. Certainly nobody believes it is wrong to deliver a baby by a Caesarean section if necessary, even though this is not a natural way of giving birth. This surgical procedure can be literally life-saving to the mothers and babies who need it. There seems to be no reason why some types of medical intervention cannot also be used to bring about conception in the first place. It is not natural, but the whole point of medical technology is to combat the negative effects of living in a fallen world and to improve health. Human beings were given the task of taking dominion over the earth, and this includes using our minds and abilities to correct problems and malfunctions, both in the world and in ourselves.

It is important to understand, however, that not every medical procedure for bringing about conception is the same. Nor should we think that just because a technology has been developed it is ethically acceptable. Furthermore, we should also recognize that couples are not under any ethical obligation to pursue medically-aided conceptions. If they have not conceived naturally, they have every right to commit themselves to God and live without children. Or they may choose to adopt. Childless couples should never feel pressured or forced to pursue medical intervention to aid with reproduction.

There are a variety of reasons why a couple may not be able to conceive. For example, some women have blockages in their fallopian tubes that prevent their eggs from getting to a place of possible fertilization and implantation. Sometimes these blockages can be removed by a simple medical operation. In this case, it would seem that receiving medical treatment to aid with fertility should be entirely noncontroversial; there is nothing wrong with it at all. In other cases, a man may have a low sperm count, or his sperm may not be mobile enough to reach the egg without assistance. Given this scenario, it is sometimes possible to collect sperm, and then to release them closer to the egg.

This procedure is known as artificial insemination. It has been used in animal breeding for a very long time. If a husband and wife decide to use the technique of artificial insemination in the hopes of conceiving a child, they are not doing anything unethical. In fact, if they make this decision, they should be supported. (It should go without saying that an inability to have children is extremely personal and can be very, very painful. Churches should provide a loving community of sensitive support, understanding, grace and compassion. Christians should also be far more careful with the way that they talk—and even joke—about the issues of marriage and children.)

Although there is nothing wrong with a husband and wife using artificial insemination in order to impregnate the woman with her husband's sperm, there is another possibility for artificial insemination that is more controversial. There are cases where a woman is inseminated with the sperm of a third-party donor. In most cases, she does not know who the donor is. Christian ethicists disagree on the ethical status of this type of artificial insemination. Some believe that it should not be done, since it introduces the genetics of a non-covenant partner into the couple's reproduction. It is possible that the husband will struggle in the knowledge that his wife provided the egg, but he was not the man who provided the sperm. Others believe that this is similar to adoption: when a child is adopted, they were not conceived or delivered by the adopting couple. Those who hold to this position maintain that the choice to care for the baby is not dependent on the genetics that produced its life, and therefore they conclude that if both husband and wife agree, it is morally permissible to use a donor's sperm. Every Christian ethicist believes that the husband's sperm should be used if possible in artificial insemination, but some are prepared to allow for donor sperm if necessary. This should be thought through carefully. (One thing that must be insisted upon is that using donor sperm in artificial insemination is not the same as committing adultery.) Couples should also ask themselves if artificial insemination involving a donor's sperm is more glorifying to God than acting to adopt a child that has already been conceived and is in need of care.

> **Every Christian ethicist believes that the husband's sperm should be used if possible in artificial insemination.**

More controversial yet is the practice of *in vitro fertilization* (IVF). In IVF, sperm and eggs are harvested and then brought together outside of the woman's body. If the sperm and eggs unite successfully, they can then be implanted into the body of the mother. IVF is possible with eggs and sperm from a wife and husband; or with either the eggs or the sperm being donated by a third party; or with neither the eggs nor the sperm coming from the married couple. Since sexual intercourse does not take place in any type of IVF, it cannot be considered adulterous. Objections to the practice must, therefore, run on other lines.

Those who support any of these kinds of IVF will suggest that the latter is like adopting a baby from their moment of conception onward. If a couple can adopt a child once it has been born, but its life begins at conception, why can't the couple care for and nurture their adopted child during the time that they are in the womb? Those who endorse IVF—in any form—must likewise maintain that the physical location of the conception is irrelevant. Whether the conception occurs in the mother's body, or if the embryo is created outside of the mother's body and is then transplanted into the mother, seems to some to be morally irrelevant. It is the sanctity of what is created, rather than where the creation occurs, that is significant. Even if we disagree with the practice (or with some forms of it), we must recognize that any human being who is conceived as a result of IVF is created in the image of God, and they are fully human. As a result, they ought to be loved and valued as much as anyone else.

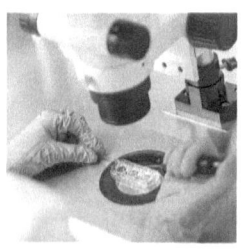

*...any human being who is conceived as a result of IVF is created in the image of God, and they are fully human.*

When it comes to IVF—even if one gives it a full endorsement theoretically—it is necessary to distinguish between IVF in theory and in practice. Theoretically, one egg could be harvested and one embryo could be created for implantation. But given the invasive nature of the procedure, as well as its high cost, when IVF is done there are usually multiple embryos that are created. Not all of these embryos are expected to be implanted in the mother. This means that embryos—human lives—are created with no intention of allowing them to survive.

Excess embryos are merely destroyed. (Some are frozen for future use, but they cannot remain frozen forever without dying.) Even if a Christian believes that IVF is morally acceptable, they need to insist that life begins at conception, and therefore that the disposal of living human embryos is morally wrong. It is not acceptable to produce several deaths in the hope of creating one growing baby. If multiple embryos are being created and destroyed in IVF laboratories, then Christians must judge the current practice as unethical. The Bible clearly teaches the sanctity of life, and this biblical principle must be upheld. (See our chapter on abortion for the biblical argument.)

## Contraceptives and birth control

There is no moral imperative for couples to have as many children as they possibly can. In the Roman Catholic Church tradition, contraceptives were considered sinful, since sexual intercourse was given for the purpose of procreation. (More will be said about the purpose of sex in our chapter on sexual ethics.) In Protestant churches, there has been far more openness to the legitimacy of contraceptives. A few points should be kept in mind:

*1. There are different types of contraceptives.*
Some contraceptives place a barrier between the sperm and egg so that fertilization cannot occur. Others place a barrier between the uterine lining and the fertilized egg so that it cannot implant and develop. In our judgement, there is nothing wrong with barriers of the first kind, but barriers of the second kind must be strictly avoided, since they work *after* conception has occurred. Intrauterine devices (IUDs) are growing in popularity, and although they are designed to prevent fertilization from occurring, they also act to prevent implantation if fertilization does take place.

*2. There are medicines that regulate a woman's fertility cycles.*
Often referred to simply as "the pill," some medications are designed to keep the woman's body from releasing eggs. This prevents conception from occurring. Some also function to induce the flushing of the uterine lining, which results in the destruction of the embryo if conception has taken place. This latter function is morally wrong. Even taking the pill requires careful thought—not necessarily because it is

a contraceptive, but because we need to be very careful about introducing medicines that are designed to affect our hormones and natural biorhythms. No medicine should be taken uncritically, and it should not be assumed that even a "safe" medicine is good to have in our bodies for an extended period of time. Couples should do proper research in order to make an informed decision about the pill, if it is an available option for them.

*3. There are permanent methods of birth control.*
Two of the most common permanent methods of birth control are surgical: vasectomies for men and tubal ligation or cauterization for women. There are some Christians who believe that any form of permanent birth control is wrong—they believe that it is sinful to surgically disrupt bodily systems that are actually functioning the way God designed them. They allow surgery to correct dysfunction, but they believe it is unacceptable to prevent the body from discharging its designed purposes. It is important for the couple to recognize that medical intervention is supposed to restore healthy and natural function. When we purposely disrupt our natural functions, we must acknowledge that there can be unintended, negative repercussions. Pragmatic factors can also be considered (eg. the possibility of complications, or potential psychological effects). We know many people who have opted for permanent sterilization but then have tried afterward to have their condition reversed so that they could have more children. Those who opt for permanent sterilization should not rush into the decision, and they should try to think about the issue theologically rather than uncritically accepting their culture's norms. We are not saying that permanent sterilization is necessarily wrong, but we are saying that it is a more serious ethical issue than many people seem to assume. Since it is not always reversible even with medical technology, it is a major decision to make. Surgical intervention that disrupts natural and healthy bodily function is a very serious matter.

*4. There are natural methods for family planning.*
Some natural family planning methods are more sophisticated than others, but they all share the principle that sexual intercourse should be avoided during the small number of days per month when the woman is fertile. The theory is that determining when a woman is

fertile can be done with charting, fluid examinations and measuring basal body temperature. There is a great deal of disagreement over how effective such natural methods are, but in principle there is nothing ethically objectionable about them. Since no unnatural devices, medicines or medical procedures are required for natural family planning, it is a completely acceptable option.

In our opinion, Christian couples have liberty—within certain parameters—to use contraceptives. For example, it is essential that these contraceptives do not result in the destruction of a fertilized egg. On the positive side, the more natural and temporary the method of birth control is, the less morally controversial it will be. We should aim to live in harmony with the way that God has designed us, and we ought to be extremely careful about medical intervention that blocks natural and healthy functioning.

Having said this, there are a variety of circumstances which may cause different couples to make different decisions about birth control. Some couples have selected permanent sterilization because the woman has health complications that make pregnancy dangerous or even life-threatening. Many couples in areas with easy access to prescription medications use contraceptive pills for a period of time, but then stop so that they can have children. Although this can be debated, it doesn't seem that short-term use of such medication is morally problematic. (It still may not be wise, however, to use any medication that isn't absolutely necessary.) We do not believe that every couple must have as many children as biologically possible. Responsible stewardship of our own fertility can be one way that we honour the Lord and exercise careful dominion over the earth. Using contraceptives does not necessarily point to a lack of faith in God's sovereignty, or to selfishness.

> ...there are a variety of circumstances which may cause different couples to make different decisions about birth control.

Regardless of the decision that is made—and although there may be a range of ethically acceptable options, not all options are necessarily equally wise or good—it is absolutely necessary for the couple to support each other in the area of reproductive health. The burden of contraception must not be unfairly shifted to one partner. Both partners must consider and honour the other. Nobody should feel bullied or forced into a decision that makes them uncomfortable, especially if the decision requires medical intervention or unnatural means. Physical, emotional and psychological support must be mutually shared. Writing about a different ethical issue, Paul said, "But whoever has doubts is condemned if they eat, because their eating is not from faith; and everything that does not come from faith is sin" (Romans 14:23). This principle is important: If one of the partners is not able to accept a proposed method of birth control in faith and with a good conscience, the other partner must not insist on that method.

## GOD'S GOOD GIFTS

A life devoted to loving God and loving others is a gift of grace. This love should flow in, through, around, to and out of the people of God. Some will love and be loved all of their lives without marrying. Others will give love in the context of a marriage covenant. Tragically, some will give love in marriage and be met by betrayal and divorce. Yet, if they are part of the family of God, they should always be the recipients of great love. Some couples will love and be loved by their children. Other couples will be married but will not have children of their own; some will adopt, but others will choose not to do so.[1] It is unspeakably heartbreaking that some children will never be loved by their parents. If the church doesn't love them, where will they ever experience love? Blood relatives ought to love one another. Neighbours and friends and community ought to be bound together in love. Because of sin we see marriages without love and parents without love and families without love—in fact, we see a *world* without love. But through the infinite redeeming merit of Christ, love is being poured out. It is not marital status that confers blessing, it is love. Jesus says that we are all neighbours, and therefore we are all to love one another. Far, far better to be unmarried and to give and receive love than to be married and unloved or

> **...each person is called to live in slightly different relational configurations, but we are all called to love.**

---

[1] We discuss the beauty and moral significance of adoption in our chapter on abortion. It is an option that mirrors the heart of God.

unloving. Jesus and Paul could be happily single because of love. God's good gifts come in a variety of forms, and each person is called to live in slightly different relational configurations, but we are all called to love. When the church loves as it ought to love, people will find far more fulfilment there than they could possibly have imagined. Family is the good gift of God, and *permanent* family is only found under God the Father, through Christ the elder brother, among the redeemed brothers and sisters who are adopted into God's family.

## *REFLECTION QUESTIONS*

1. Does your church have a healthy and biblically balanced view of marriage, families and singleness?

2. If divorce is permissible in certain circumstances, is remarriage after divorce permissible in those cases as well?

3. What are other Bible passages that deal with the topic of infertility and conception?

4. Are there areas of child abuse in your community that you need to actively oppose? What are you doing? What should you start doing?

# 5

# Sexual ethics

The human race would not exist, and could not continue to exist, without sex. This one simple fact proves that sexual issues are relevant at all times and places, as well as in every society and culture around the world. Yet, different cultures can have very different degrees of openness when it comes to discussing sex and sexuality. There have been times in Western culture when sex was a topic that was kept relatively quiet and undiscussed. Today, however, Western culture is extremely explicit and open about sex—even young children are taught about it in many of our schools. We recognize that we face a challenge when it comes to addressing the matter of sexual ethics in a book that is designed for global distribution. We understand that different readers in different cultures will have varying levels of comfort about these discussions. Nobody can escape being influenced by their society, and in the West sexuality is so exploited and flaunted that it is difficult to remember how unhealthy and abnormal these views of sex and sexuality actually are. It is tragic that North America's debased view of sexuality—as displayed in advertising, entertainment, social philosophy and practice—is being exported around the world.

Western society has tried to simultaneously hold two contradictory views of sex. First, sex is seen as incredibly special, the highest pleasure that must be experienced at all costs; it may even be considered a transcendent experience. In this view, sex has been virtually deified. Second, sex is seen as merely a biological function to satisfy a biological drive. It is lowered to nothing more than an animal act, no different than digesting food or scratching an itch. In this view, sex is nothing special at all. So Western society now approaches sex paradoxically as something that both is and is not of any special significance. Regardless, the one clear fact is that sex is an obsession in Western society.

The freedom to have sex without reproductive consequences is one of the main feminist arguments for legalized abortion. However, when did engaging in sexual behaviour become a right that is detached from its natural consequences? Trying to separate sex from responsibility has been disastrous. The consequences of irresponsible sexual behaviour can include broken homes, divorce, guilt, rejection, shame, unwanted pregnancy, abortion, disease and death. Lest it seem like death is an extreme item to add to the list, we need to remember the number of people who have contracted HIV/AIDS from sexual encounters. I, Steve, have been in an AIDS hospice in Africa, visiting people who were dying from a disease they acquired through sexual intercourse. Make no mistake: sexual ethics can be a matter of literal life and death.

## SEXUAL MORALITY

If we are to understand what is sexually unethical, we must understand the nature of sex and sexuality. This requires, of course, a return to worldview-level thinking. The nature of human beings is of crucial importance for arriving at proper conclusions about sexual ethics. We must also think about the existence and nature of God. If we are created by God, then we are not autonomous: our bodies belong to God, as does everything else in creation (1 Corinthians 6:19). We are stewards of our bodies, so we are responsible to do with them what honours him (rather than just doing whatever we feel like or desire). If there is no God and there is no point to human existence, then it would seem that there is no reason to have any sexual ethics at all (or any ethics whatsoever, for that matter).

If human beings are the accidental products of material evolution, then sex is merely an animal function. Interestingly, at this point it could be argued that it is logically possible that some atheists have made a terrible error in rejecting monogamy in their sexual ethics. If the atheistic story of naturalistic evolution is true, then it is possible that we have evolved to flourish when sex is restricted to monogamous, lifelong relationships. We may be psychologically and emotionally adapted for sexual bonding, which then means it is harmful to be sexually promiscuous. The atheist may be reducing sex to an animal function alone, but this is an entirely inadequate guide to the complexity of human experience. (After all, they speak against rape and other forms of sexual and non-sexual violence, but such things can be reduced to merely animal functions as well.) It could be that, when considered holistically, even the atheist should warn against promiscuous sexual experiences. How does the atheist actually know that their prescriptions are beneficial rather than harmful, when considered in terms of net gain? It is not a necessary conclusion that evolving in a materialistic world means that we have evolved to be psychologically and emotionally healthy outside of sexual monogamy.

As a Christian, sexual matters are seen through the lens of the biblical worldview. There are a number of preliminary considerations which are important to observe. First, since God is the creator of everything, God is the one who created humans as sexual beings. Sexuality and sexual intercourse were part of God's original creation design, which he determined was very good. Thus, sex must be understood as something which was created by God as a good gift to the created order. Second, sexual intercourse was designed to accomplish the biological function of reproduction. God could have decreed that children be conceived in any number of ways, but he chose sexual intercourse as the means of generating new life. Third, God restricted sexual intercourse to a man and a woman who are partners in the covenant of marriage. This is not merely for the purpose of keeping reproduction within a certain covenant frame. Sexual intercourse does have a reproductive function, but it is a mistake to reduce it to only having a reproductive function. It is

> ...sex must be understood as something which was created by God as a good gift to the created order.

also a means of bonding and experiencing pleasure. Christian ethics will not minimize God's intention in this regard.

## Biblical texts on the goodness of sex

### 1. Genesis 2:24-25

> That is why a man leaves his father and mother and is united to his wife, and they become one flesh. Adam and his wife were both naked, and they felt no shame.

In marriage, a man and a woman form a new primary relationship, and this relationship is bonded by becoming one flesh. Adam and Eve were not merely nude, they were emotionally transparent (before sin there was nothing to hide, externally or internally). This is a clear statement that marriage and sex were designed by God as part of the human experience before sin entered the world. Original sin, therefore, was not sexual intercourse, as some have mistakenly believed.

### 2. Hebrews 13:4

> Marriage should be honored by all, and the marriage bed kept pure, for God will judge the adulterer and all the sexually immoral.

This text is obviously a warning against sexual immorality and illicit sexual intercourse, but the reason for the warning is grounded in the positive reality of the holiness and purity of sex in its proper context. Christians need to remember that the Bible's sexual ethic is positive toward proper sexual expression and experience. It is a joy and delight. Sin, however, distorts and twists good gifts. This is why sexual ethics are so important: sex is special, and thus has the potential to be a great blessing, or to result in great loss, harm and judgement.

### 3. Song of Songs 7:1-13

> How beautiful your sandaled feet,
>   O prince's daughter!

Your graceful legs are like jewels,
   the work of an artist's hands.
Your navel is a rounded goblet
   that never lacks blended wine.
Your waist is a mound of wheat
   encircled by lilies.
Your breasts are like two fawns,
   like twin fawns of a gazelle.
Your neck is like an ivory tower.
Your eyes are the pools of Heshbon
   by the gate of Bath Rabbim.
Your nose is like the tower of Lebanon
   looking toward Damascus.
Your head crowns you like Mount Carmel.
   Your hair is like royal tapestry;
   the king is held captive by its tresses.
How beautiful you are and how pleasing,
   my love, with your delights!
Your stature is like that of the palm,
   and your breasts like clusters of fruit.
I said, "I will climb the palm tree;
   I will take hold of its fruit."
May your breasts be like clusters of grapes on the vine,
   the fragrance of your breath like apples,
   and your mouth like the best wine.
May the wine go straight to my beloved,
   flowing gently over lips and teeth.
I belong to my beloved,
   and his desire is for me.
Come, my beloved, let us go to the countryside,
   let us spend the night in the villages.
Let us go early to the vineyards
   to see if the vines have budded,
if their blossoms have opened,
   and if the pomegranates are in bloom—
   there I will give you my love.
The mandrakes send out their fragrance,
   and at our door is every delicacy,

both new and old,
  that I have stored up for you, my beloved.

Although some of the imagery may not pass directly from one culture and time to another, this passage connects the language of poetry and metaphor to the physical description of the woman's body. The man delights in the woman's physicality and beauty, and she enjoys his delight. In other passages in Song of Songs, the woman likewise expresses her delight in her lover's body. The passage cited above is very sexually explicit. A study of the Hebrew metaphors reveals how *extremely* explicit many of the images are (an explicitness which is not naturally perceived by us, since we don't use the same metaphors in our sexual vocabulary). Yet, even apart from studying the metaphors, there can be little doubt about the content of verses 7–8. As a palm tree has a slender trunk that leads up to its clusters of fruit, and as climbing the tree requires wrapping around it, so the man wraps himself around the woman, and takes hold of the fruit. We ought not to be more prudish than God. We must also, however, not be foolish and lower sexual experience to the level of the commonplace and trivial. There ought to be a holy and reverent veil over personal sexual experience, but this veiling must be balanced by honesty in discussion. The Bible has a perfect balance: it is open, honest and transparent about sex, but it is also reverent and discreet. It is not a closed topic, but neither is it a topic that is shared without decorum.

Sex, then, is to be the consummation of commitment that is solidified in a marriage covenant. It is designed to bring the husband and wife into ever-increasing levels of intimacy (i.e., oneness). Ideally, in marriage, ongoing sexual experience is located in the safe and loving relationship of acceptance. Sex should not be something that one tries to *get*, it is something that one should try to *give*. One ought to consider the marriage partner rather than oneself. Sexual intercourse should be an expression of mutual love and concern, where the focus of each person is on their partner. When this takes place inside of a marriage, there is security. Sex is not about physical performance: it is about sharing love. First Corinthians 7:1–7 talks about the mutuality and giving that should characterize marital sex. Husbands and wives are to put one another first. In this sense, sex is no different than anything else in marriage—selfishness is *always* to be excluded, no matter what the subject! Husbands and wives should each view their body as being given for the good of the other, and they should each place their own desires under what's best for their spouse.

Since the sex drive is a very powerful force, and sexual behaviour has very powerful physical, mental, emotional and spiritual consequences (for good or for ill), a society's sexual practices will have an enormous impact on both individual and community health. Whereas sex in proper context can provide for positive bonding and relational growth, when it is misused in the wrong context, people suffer.

> **Husbands and wives should each view their body as being given for the good of the other, and they should each place their own desires under what's best for their spouse.**

Proper sexual behaviour strengthens and protects the community. Improper sexual behaviour weakens and harms the community. Sexual energy needs to be properly channelled, for the sake of the individual but also for the sake of society. One does not love their neighbours if they exploit or harm them sexually—and this harm can be inflicted even if the experience is consensual. The Bible's view of sex in its proper context is that it is holy, pure and good. Wresting it out of its proper context, therefore, is extremely damaging. It is also wicked.

## SEXUAL IMMORALITY

The potential for the gift of sex to be abused needs to be taken very seriously. Unfortunately, this potential is being actualized: sexual immorality is rampant in the world today. In the Western world, violations of God's law are actually celebrated. Recently a young man in our city who is looking to enter pastoral ministry shared his testimony with a group of pastors. He said that he had been involved in a life of hedonism, living for pleasure. He said that it left him empty (this is a common experience, but one that our media and world neglect to mention). Then he said something profound. He said that in retrospect, he hadn't been looking for hedonistic pleasure, he had been looking for community and relationship. That insight is worth remembering and pondering. What people truly need is love and community. That idea comes up numerous times in this book.

One of the terrible and tragic results of sexual immorality is that it breaks down genuine intimacy, increases loneliness and *severs* rather than *establishes* connection. Sexual intercourse is designed as an implicit declaration of lifelong commitment, so when it is experienced casually, it is not surprising that a feeling of betrayal and isolation can be generated. Sometimes people are quite conscious of this reality; others simply experience a vague sense of unease and emotional or spiritual deadening. We were not designed to experience abandonment by a sexual partner. It is unethical, therefore, to engage in sexual activity with anyone to whom one is not committed for life. This commitment, furthermore, needs to be sealed in a marriage covenant before it is assured. Many people have had promises of commitment before marriage that have led to sexual conduct, only to find that the marriage never takes place. In many countries, it is illegal to drive without a driver's license—driving a car with the excuse that one is going to get a license tomorrow is not valid in the eyes of the law. Similarly, the promise of marriage tomorrow does not validate sexual intercourse today.

> **One of the tragic results of sexual immorality is that it breaks down genuine intimacy, increases loneliness and severs rather than establishes connection.**

The sheer number of references in both the Old and New Testaments to sexually immoral behaviour helps us see how common and important the topic is. Fornication, adultery, bestiality and incest are all expressly and categorically forbidden. There is simply no argument or debate about this point. At one level of analysis, we can simply say that God has given his law and revealed his moral will very clearly when it comes to sex—it is our job to obey. Nothing else needs to be known than that God has spoken with authority and we are morally responsible to him for our actions. Ultimately, the reason we obey God's commands is to give him glory. Obedience is also for our good; it is the only way we can flourish. We must trust that God is keeping us from harm, and that he has given us these instructions for our good.

## 1 Corinthians 6:12-20
A key text for understanding sexual immorality is found in 1 Corinthians 6:12–20. Paul writes:

> "I have the right to do anything," you say—but not everything is beneficial. "I have the right to do anything"—but I will not be mastered by anything. You say, "Food for the stomach and the stomach for food, and God will destroy them both." The body, however, is not meant for sexual immorality but for the Lord, and the Lord for the body. By his power God raised the Lord from the dead, and he will raise us also. Do you not know that your bodies are members of Christ himself? Shall I then take the members of Christ and unite them with a prostitute? Never! Do you not know that he who unites himself with a prostitute is one with her in body? For it is said, "The two will become one flesh." But whoever is united with the Lord is one with him in spirit.
>
> Flee from sexual immorality. All other sins a person commits are outside the body, but whoever sins sexually, sins against their own body. Do you not know that your bodies are temples of the Holy Spirit, who is in you, whom you have received from God? You are not your own; you were bought at a price. Therefore honor God with your bodies.

In this passage, Paul repeats slogans that some in Corinth were using. When they say, "Food for the stomach and the stomach for food," they are using an *entendre* (i.e., a statement with a double-meaning) to say that our bodies were made for sex, therefore having sex whenever we feel like it—with whoever we feel like it—is natural and acceptable, just like eating food. But notice that Paul corrects their assumption that their body belongs to *them*. It doesn't. Their body belongs to the Lord. Our bodies will be resurrected, and this shows how highly valued they are as God's property. Paul argues that if we are united with Christ, and we unite ourselves to a prostitute, we are defiling Christ (in a manner of speaking). This same principle would apply to all unlawful sexual intercourse, whether it be fornication or adultery. Paul's advice is clear: Flee from sexual immorality! This is the same attitude adopted by Joseph with Potiphar's wife in Genesis 39, when he literally ran out of the room where she was trying to seduce him.

Paul goes on to provide additional reasons for fleeing sexual immorality. When we sin sexually, we are sinning against our own body. In illicit sexual intercourse, we are uniting with someone who will fragment us. (We are also harming others, and this is always unethical; it violates the principle of love.) If we engage in sexual immorality, we are engaging in self-harm. Given the enormous number of sexually transmitted diseases in the world today, one cannot help but see how this principle applies both spiritually *and* physically. Believers in Christ are one with him and are also the temples of the Holy Spirit. Does the Spirit want his temple marked by sexual defilement? The answer is obvious. For the seriousness of this analogy, one needs to read the Book of Ezekiel and other Old Testament prophets.

Paul's last appeal is to our redemption. Christ bought us with the price of his own blood. We belong to him. Therefore, we must honour God with our bodies. Sexual immorality is simply forbidden, and Christians need to flee from it in every form.

### *Adultery*

Adultery is sexual sin that violates an existing marital covenant when one partner in the marriage has sexual relations with someone other than their spouse. It is a subset of the wider category of sexual immorality, and it is clearly evil in the sight of God. Jesus gives adultery as grounds for divorce,[1] so it is extremely serious in God's sight. The reason we are discussing adultery is not because the Bible's teaching is unclear, but rather because it provides a lens into how sexual ethics is approached from both a positive and negative direction in Scripture.

### *1. Exodus 20:14*

"You shall not commit adultery." This text is not lacking in clarity. It is a categorical rejection of the permissibility of adultery. Sometimes Christians in an unhappy marriage have said that they know God wants them to be happy, so God would not object to them forming a sexual relationship with someone other than their spouse. This reasoning is self-serving, dishonest and unbiblical. Adultery is never an ethically allowable option.

---

[1] Matthew 5:27–28. For further discussion, see Chapter 4.

## 2. The Prophets

Various texts could be cited, but the prophets often compare religious idolatry with marital adultery. In idolatry, the people are being unfaithful to God, and the closest analogue the prophets can find is with adultery. This shows how wicked adultery is in God's sight.

## 3. Genesis 39:9

"My master has withheld nothing from me except you, because you are his wife. How then could I do such a wicked thing and sin against God?" Joseph, when given the opportunity to have sex with a woman who had power over him in his master's household, fled because sleeping with her would be a wicked thing and a sin against God. It is vital to remember that adultery is not only a sin against one's marital partner and a sin against the person with whom one is committing adultery, it is a sin against *God* (cf. Psalm 51:4).

## 4. Proverbs 5:1-23

(Only selected verses are reproduced below.)

> For the lips of the adulterous woman drip honey,
>     and her speech is smoother than oil;
> but in the end she is bitter as gall,
>     sharp as a double-edged sword.
> Her feet go down to death;
>     her steps lead straight to the grave.
> She gives no thought to the way of life;
>     her paths wander aimlessly, but she does not know it.
>
> Keep to a path far from her,
>     do not go near the door of her house,
> lest you lose your honor to others
>     and your dignity to one who is cruel,
> lest strangers feast on your wealth
>     and your toil enrich the house of another.
> At the end of your life you will groan,
>     when your flesh and body are spent.

> May your fountain be blessed,
>   and may you rejoice in the wife of your youth.
> A loving doe, a graceful deer—
>   may her breasts satisfy you always,
>   may you ever be intoxicated with her love.
> Why, my son, be intoxicated with another man's wife?
>   Why embrace the bosom of a wayward woman?
> For your ways are in full view of the Lord,
>   and he examines all your paths.

This is an important text. The reasons for avoiding adultery are moral, spiritual and pragmatic. Committing adultery brings ruin, shame and disgrace. The antidote to adultery is also provided in this text—those who are married ought to seek sexual satisfaction *in* their marriage. (If this is impossible due to illness or other difficulties, spouses are still not allowed to find sexual outlets outside of their marriage. Forced marital celibacy is no different than the required celibacy of singleness.) Just like the Song of Songs, we find that the sexual language is explicit, but not crass. The Book of Proverbs has a different style in chapters 1 to 9 (block discourses) than chapters 10 to 30 (pithy sayings). A careful reading of chapters 1 to 9 shows a concern with adultery that comes through again and again. For example, 6:25–35 is another passage that lays out the consequences for adultery in no uncertain terms. The amount of space given to the topic of adultery in a book devoted to wise living shows how foolish those who engage in adultery are. Wisdom requires covenantal fidelity.

## 5. Romans 13:8-10

> Let no debt remain outstanding, except the continuing debt to love one another, for whoever loves others has fulfilled the law. The commandments, "You shall not commit adultery," "You shall not murder," "You shall not steal," "You shall not covet," and whatever other command there may be, are summed up in this one command: "Love your neighbor as yourself." Love does no harm to a neighbor. Therefore love is the fulfillment of the law.

Notice that this text says that love fulfils the law, and therefore love will not permit adultery. It stands to reason that if someone loves their spouse, the possibility of being tempted to commit adultery diminishes. Temptation, however, is real, because nobody loves perfectly. People sometimes give in to sin against their better knowledge and judgement. There are times when we simply fail through a weakness of will. If we were perfected in love, we would not sin in any way. Often when people commit adultery, they justify it by saying that they love the person with whom they are having the adulterous relationship. But God says otherwise. If you *truly* love someone, that love will keep you from committing adultery with them. Love does no harm, and since adultery harms everyone involved, adultery is proof of inadequate love. The more one actually loves, the less adultery becomes possible. The antidote to adultery isn't being apathetic toward those who aren't one's spouse; it is loving them. The danger here, of course, is that we don't love perfectly and we are susceptible to temptation and failure. We must not confuse feelings with biblical love. Biblical love prevents, not causes, adultery. Living in a fallen world requires conducting all relationships with wisdom, goodness, virtue and love. We must love someone enough to care for their honour forever, rather than for our own pleasure now. That is a mark of true love.

> **We must love someone enough to care for their honour forever, rather than for our own pleasure now.**

### Lust

Currently, the Western world does virtually all that it can to incite lustful thoughts and feelings. Advertisers know "sex sells." The entertainment industry would almost cease to exist if every show, movie and song containing sexually inappropriate material was banned. In such a society, the words of Jesus could not be more relevant: "You have heard that it was said, 'You shall not commit adultery.' But I tell you that anyone who looks at a woman lustfully has already committed adultery with her in his heart" (Matthew 5:27–28). There is one vitally important principle at work here: If you do not *lust*, you will not end up committing fornication or adultery. Lust is a beginning step toward sexual acts.

Those who have self-control when lustful feelings or thoughts arise will have more power to avoid being swept away when a more serious temptation presents itself. Jesus is telling us that sin begins in our *heart*. What happens in one's heart and mind makes external actions either more or less possible. Where there is sexual impurity in action on the outside, it is a guarantee that there is sexual impurity in the intention of the heart and mind. Lustful thoughts may seem private, but they have a way of manifesting themselves in external acts. Even the semblance of privacy is illusory, of course. God always knows what we are thinking. More than this, however, how you think cannot help but affect how you treat others and interact with them.

**When someone views pornography, they are completely destroying the connection between commitment, relationship and sex.**

Pornography has never been necessary for generating lust; people have lusted without ever seeing pornographic materials. But the entire pornographic industry is designed with only one purpose—it is designed to make money by producing lust in its consumers. Pornography is an abomination in our world. It objectifies and dehumanizes those who participate. Consumers of pornography are reduced to economic units to be exploited. Those who perform in pornography are by definition engaged in sin. When someone views pornography, they are completely destroying the connection between commitment, relationship and sex. The person they are lusting for has been reduced to a picture or event; they have become unhuman. It is not a *person* who is being viewed; the person has been reduced to a body or movement. And the person who is the object of the viewer's lust does not know the viewer. Presumably they know that someone will be viewing them, but they do not know who or when or how. There is no mutual exchange, no companionship, no warmth, no love. There is only selfishness, exploitation and dehumanizing objectification. A stranger on a computer screen cannot be a partner in covenantal sexual expression.

Viewing pornography on a consistent basis alters brain connections in regards to sex. What should be a bonding experience now becomes an isolated and isolating one. Sex is communication, but there is no communication in pornography. People begin to equate sex with selfish, lonely pleasure. In trying to be fed, people who look at pornography guarantee that they will suffer from sexual malnourishment. Marriages have been ruined by pornography, and many who have entered marriages have found themselves severely handicapped and immature in sexual expression, having fallen into the deceptions of the pornography industry. When pornography shapes how people view sex, it is always destructive. When people are habituated to lust, it changes how they look at and perceive other people. When everyone is seen as a possible sex object, it changes how people talk to each other, look at each other and even touch each other non-sexually.

**When people are habituated to lust, it changes how they look at and perceive other people.**

It is a tragedy that girls and women need to live in a world that is swamped by internet pornography. It is a tragedy that boys and men live in a world where access to something so destructive is almost always present and instantaneously available. Christians must oppose this industry and must *never* support it by viewing its products. If the makers of pornography could not find a market of buyers, the industry would disappear. Pornography is made because it is financially lucrative for those who produce it. We need to hear and obey Jesus' words, and root out all lust from our hearts.

Imagine a world where people acted in line with Paul's words to Timothy: "Do not rebuke an older man harshly, but exhort him as if he were your father. Treat younger men as brothers, older women as mothers, and younger women as sisters, with absolute purity" (1 Timothy 5:1–2). In regards to lust, it is the latter category of younger women that is so important. When women are around men, they should not feel that they are being looked at as merely sexual objects. Christian men should treat younger women like sisters in Christ. They can recognize that their sisters are beautiful and attractive, but they must not lust and objectify them as sexual objects. Treating a younger woman as a sister would make her feel loved, valued, respected and honoured, *not* objectified or unclean. There is to be absolute sexual purity in these relationships. A brother who loves his sister always treats her as an equal person. She should know it and feel it. She is protected and safe in the community because her brother in Christ will never harm her. After all, he loves her; she is his sister. And, in Christ, she loves her brother, too. She will take steps to guard her brother's heart and mind. She will not dress seductively or attempt to ensnare men by flaunting her sexuality. They will mutually keep each other safe.

## *Homosexuality*[2]

In a number of nations in the world today, homosexual acts are illegal. Engaging in homosexual practices can even lead to judicial execution in a handful of countries. There are many more cultures in which homosexual behaviour is not illegal, but where it is still considered immoral. In contrast, the Western world has experienced a historic shift in its view of homosexual orientation and practice. In Canada, homosexual acts were illegal until just before the end of the twentieth century. Just a few years later, Canada gave homosexual couples the right to be legally married. Many cities around the world—both large and small—have annual events where homosexuals parade through the streets with great fanfare. There is a myriad of things that some societies do that are designed to normalize and celebrate homosexuality.

So we live in a world where many nations celebrate homosexuals, and other nations execute them. While there are principles that all Christians need to understand and apply when it comes to homosexuality and ethics, we recognize that this issue will be perceived both through Scripture and our cultural situations. In some places the Bible's view of homosexual acts is shared by the wider society; in other places, the Bible's view is rejected.

---

[2] One of the difficulties in writing this section was choosing the vocabulary. In North America, it is common to refer to the LGBTQ community (LGBTQ stands for lesbian, gay, bisexual, transgender and queer). But LGBTQ is seen as far too limited by some, and they want more types of sexual identities included. The result is that the accepted vocabulary is changing all the time. Given that this book is written for an international audience, we have decided to use the simplest vocabulary possible. Unless context indicates otherwise, the term "homosexual" (and cognates) will include anyone who is same-sex attracted. There is simply no standardized vocabulary that will be coherent in every culture and language. We are aware of this limitation, but trust that our meanings will be clear.

But what does the Bible actually say? This is a topic that desperately requires proper hermeneutics and biblical interpretation. The Old Testament law reveals God's view of the morality of homosexual acts, but no nation on earth is a theocracy like ancient Israel. Without rejecting the authority of the Word of God, we see that the Word itself reveals progressive development over time. God's covenant with Israel was a national and legal covenant, which is not identical to the new covenant. Thus, a Christian can maintain that homosexual acts are immoral, yet still disagree that the Old Testament legal penalties for such behaviour should be enforced by a nation's government and legal system. Failure to observe the Sabbath Day also carried with it the penalty of death, but there are not too many Christians today who believe that every nation on earth should enforce such a prescription. Nevertheless, the uniform interpretation of the Christian church throughout its entire history has been that both the Old and New Testaments are clear that God does not approve of homosexual practices. In fact, throughout the history of Jewish and Christian interpretation, it would seem very strange to have to make that case, since the texts are overwhelmingly clear. Even non-Christians recognize that the Bible does not endorse the sexual acts of homosexuals. We will present a selection of biblical texts, with minimal commentary. Then, at the end of this section, we will add some additional considerations.

> *God's original design was for heterosexual coupling, as male and female bodies demonstrate.*

1. Genesis 1-2

God created the human race to consist of males and females. Adam and Eve were created and brought together into a heterosexual marital relationship. There is a biological function and anatomical fit with heterosexuals that is not possible in homosexual relations. Although the text does not speak of homosexuality, God's original design was for heterosexual coupling, as male and female bodies demonstrate.

2. Genesis 19

In history, the destruction of Sodom and Gomorrah has always been seen as, at least in part, a judgement on homosexual behaviour. Yes, the sins of Sodom included far more than just homosexual activity, but homosexual activity cannot be removed from the equation. In English, the meanings of the words *sodomy* and *sodomize* are linguistic proof of how Genesis 19 has been interpreted historically. There are some today who are trying to revise the historical interpretation by making the homosexual aspect irrelevant, but their arguments simply fail to deal with the details of the text.

### 3. Leviticus 18:22 (cf. Leviticus 20:13)

"Do not have sexual relations with a man as one does with a woman; that is detestable." This proscription is clear. In God's law, homosexual sexual acts are forbidden. In God's sight, they are detestable (or "an abomination"). It is worth noting that the law also forbids a great number of *heterosexual* sexual acts as well. The vast majority of sexual sin in our world is heterosexual rather than homosexual, and it is also detestable in God's sight.

### 4. Romans 1:26-27

> Because of this, God gave them over to shameful lusts. Even their women exchanged natural sexual relations for unnatural ones. In the same way the men also abandoned natural relations with women and were inflamed with lust for one another. Men committed shameful acts with other men, and received in themselves the due penalty for their error.

Despite attempts to interpret this passage in a way favourable to homosexual practice, all such efforts have failed. Gay and lesbian acts are declared to be unnatural and sinful in this text. The rest of Romans 1:18–32 should be read for the context. Paul is very clear that homosexual acts are not the *only* sexual acts that God deems to be sinful. Beyond this, Paul makes it very clear that *every* manifestation of sin comes from the same source, which is a sinful heart that rejects the knowledge of God in unrighteousness. Engaging in homosexual activity is as much the product of a sinful heart as engaging in heterosexual immorality. Or gossiping. Or being proud. Or failing to love. *Every sin* flows out of the same heart problem.

**Every sin flows out of the same heart problem.**

## 5. 1 Timothy 1:8–11

> We know that the law is good if one uses it properly. We also know that the law is made not for the righteous but for lawbreakers and rebels, the ungodly and sinful, the unholy and irreligious, for those who kill their fathers or mothers, for murderers, for the sexually immoral, for those practicing homosexuality, for slave traders and liars and perjurers—and for whatever else is contrary to the sound doctrine that conforms to the gospel concerning the glory of the blessed God, which he entrusted to me.

This is quite a list. Practicing homosexuals are on it, and the context is explicit that everything on this list is evil in the sight of God. Notice again, this list is not exhaustive in detail: Paul adds that *anything* that is not in conformity to Christ is evil. Someone may not be guilty of homosexual practice, but that hardly makes them innocent in the sight of God.

## 6. 1 Corinthians 6:9–11

> Or do you not know that wrongdoers will not inherit the kingdom of God? Do not be deceived: Neither the sexually immoral nor idolaters nor adulterers nor men who have sex with men nor thieves nor the greedy nor drunkards nor slanderers nor swindlers will inherit the kingdom of God. And that is what some of you were. But you were washed, you were sanctified, you were justified in the name of the Lord Jesus Christ and by the Spirit of our God.

The text is unequivocal in listing "men who have sex with men" in the category of wrongdoers (which puts them in the same boat, once again, with those who practice heterosexual immorality). The phrase "men who have sex with men" translates Paul's Greek, where Paul uses words that refer to both the active and the passive partner in the homosexual act. Both partners are regarded as sinning.

But this text is also vitally important for how Christians are to respond to homosexuals. Paul is not afraid to label sin as sin. He speaks the truth, but in love. However, he also says that some of the Corinthian Christians *were in the past* counted among the members of that list. But

God is a God of grace. They have repented and turned away from their sin, no longer practicing it. Their past does not define them. They were washed, so they are both clean and able to enter the covenant community. They are sanctified, so they are holy and placed in the sphere of the sacred—they are reserved only and always for the God who is "Holy, Holy, Holy." They are justified, declared legally innocent in the sight of God. In Christ, their guilt is atoned for and their sin is removed. Christ paid their penalty and died their death. Now the formerly-practising homosexual stands as a new creation, holy and pleasing to God, through Christ, by the Holy Spirit. They are caught up in pure love. They are part of the body. In Christ, redeemed homosexuals and lesbians are our brothers and sisters. And, just as we ought to love the lost regardless of their particular sin, so we ought to love those who are engaged in same-sex sexual behaviour and practices. Christ only shed his blood for those he loved, and thus if homosexuals are saved through his blood, Christ loved lost homosexuals as well as lost heterosexuals. As Christ's body, we can do no less.

*Additional considerations*
What is the relationship between sexual orientation and birth? Many today argue that homosexuals should be able to engage in homosexual acts because they were born with same-sex attraction. From the very start, we do need to reject the position that some Christians have taken, which is that people with a homosexual orientation always just make a simple choice to be homosexuals. That position is false. But neither is there strong scientific evidence to support the view that homosexuals are born with that orientation ingrained in their DNA. There is a host of sociological data which indicates that environmental factors contribute to same-sex attraction. Nevertheless, all such claims (for both heredity and environment) are challenged. Perhaps the most non-controversial position to take at this point is to say that some people may have a genetic predisposition to same-sex attraction which might be activated given certain environmental factors. (Notice the words "may" and "might" in that sentence. More data needs to be gathered, and it needs to be carefully analyzed apart from political agendas.)

What follows from this position in regards to the morality of homosexual acts? Nothing. In fact, we can go further and grant for argument's sake that homosexual orientations are hardwired into someone's genetics, and they are genetically fated to be attracted to members of their own sex. Does this fact justify engaging in homosexual behaviour? Of course not. The vast majority of people in the world have heterosexual orientations (which they did not choose, either). But does this mean that people can engage in whatever sexual activity they want? Absolutely not! The Bible is fairly clear on that point. Fornication is forbidden. If a married woman finds herself desiring a sexual relationship with a man who is not her husband, can she justify adultery on the basis of being born a heterosexual? Jesus forbids *lust*. We are to be in control of our bodies, and that includes our minds, not just our sex organs. There are also many Christians who have healthy and normal heterosexual desires, and strong sex drives, but they are unmarried and thus cannot have sexual intercourse without falling outside of God's law. What are they to do? They are to remain faithful to God and exercise self-control. The unmarried heterosexual is no more able to engage in holy sexual acts than are homosexuals.

The objection has been made that at least heterosexuals can get married. While this is technically true, there are many Christians who are not married and who will never marry, and this is not always by choice. They will have to practice lifelong sexual abstinence. In fact, there are far more single Christians with a strong heterosexual sex drive who will never marry than the number of redeemed Christians who struggle with same-sex attraction, and both groups are expected to abstain from sexual relations. If people are unmarried, God expects them to practice sexual abstinence, whether their desires are heterosexual or homosexual. Being tempted toward sexual sin is not itself sinful. Jesus was tempted by Satan but did not give in to sin. Someone may be tempted heterosexually or homosexually, but this by itself does not mean they are sinning.

**If people are unmarried, God expects them to practice sexual abstinence, whether their desires are heterosexual or homosexual.**

Part of our problem in the Western world is that we have an entirely distorted view of the nature and importance of sex. In the Christian subculture, we also tend to have a distorted view of the nature and importance of marriage (see our chapter on family ethics). Sadly, many people in our society cannot believe that a life without active sexual behaviour can be fulfilling. But that perspective is entirely wrong. Jesus was unmarried and celibate. Singleness can be a great gift of God's grace. Even those who will one day be married need to practice years of self-control and sexual abstinence. The church needs to speak a powerful and counter-cultural message, maintaining by both word and deed that sex is a gift from God, but not the *highest* gift from God. A life without sex can be completely fulfilling. After all, the steadfast love of the Lord is better than life (Psalm 63:3). It is also—as many have tragically discovered—a fact that many people who are having lots of sexual experiences are lonely and unfulfilled.

The church needs to be a place where people who are trying to honour God are loved rather than condemned. There are many believers who were practicing homosexuals before they were saved. Such a person may be saved, yet still find themselves attracted to the same sex (just like a heterosexual person addicted to pornography can be saved and yet struggle with that temptation). Having a homosexual orientation does not remove someone from God's saving grace. The church needs to be a place where nobody is shamed or made to feel small on the basis of their orientation. We have known Christians who struggled with same-sex attraction all of their lives—they certainly didn't need others to make them feel guilty or ashamed. No, the church needs to provide compassion, empathy and love to all—especially to those who are struggling.

**The church needs to be a place where people who are trying to honour God are loved rather than condemned.**

Once again we return to a common theme of this book—*everyone* needs love and community. Sex is not something that everyone needs. Is self-control and sexual abstinence easy? You can answer that question for yourself. But love is infinitely more important. How heartbreaking it is that so much of the sexual activity in our world is completely selfish and unloving. Is an act of heterosexual intercourse where one person is consciously and crassly using the other person—only to discard them the next morning—less offensive to God than a homosexual encounter between two people who are truly friends and who give in to temptation in a moment of weakness? Asking this question does not justify any homosexual practice—but it does raise the issue of the relationship of feelings and intentions in sexual acts, rather than just focusing on the act itself. It is entirely possible that some heterosexual acts are worse than homosexual acts, especially when they are spiteful, selfish, lacking in compassion or even hateful. Sex is not our saviour, and our Saviour never experienced sex. He offers us something better—he offers us love.

### Prostitution and sex slavery

Given the Bible's general sexual ethic, it is logically impossible to justify prostitution. Paul emphatically states, "Shall I then take the members of Christ and unite them with a prostitute? Never!" (1 Corinthians 6:15). Hiring a prostitute is abuse. It is not only the sexual act which is sinful, it is the context. The prostitute is reduced to a body—they are certainly not respected as an individual person. They are dehumanized and reduced to their body parts, and what their body parts can do to the client's body parts. There is also no denying the fact that prostitutes have often fled abusive situations (frequently ones where they have been sexually molested as children). They often turn to alcohol and drugs to numb their feelings. Sex is supposed to heighten feelings, but it deadens the heart of the prostitute. Many who turn to prostitution do so out of a sense of abject hopelessness. The church needs to offer them hope. If the church does not see the *person* behind the prostitute, who will? Prostitutes can get money for sex, but where will they get genuine love?

Even worse, large numbers of prostitutes are either officially or unofficially slaves. Around the world, prostitution rings are being controlled by fear, drugs, violence, rape and murder. The human cost of this is beyond calculation. Christian missions and organizations **Sex slavery is one of the major social justice issues of our day.** have been established to fight against such realities, but there is far, far more to be done. Sex slavery is one of the major social justice issues of our day. Christians cannot afford to ignore it, and affluent churches must not hoard their resources. We are thankful to God that our home church has partnered with agencies to help fight child sex slavery in the

Philippines, both by prosecuting the offending adults and by rescuing and restoring the child victims. Our contribution is just the tiniest drop in the ocean, but it is better than nothing. If we can even save one life from this nightmare, it is worthwhile. Adult prostitution is bad enough, but when children are forced into performing sexual acts so that adults can make money, we have reached the nadir of depravity. It is an abomination that there is a market for this. Paying to use a child for one's own sexual gratification is an evil beyond description.[3]

Victims of prostitution—both children and adults—need to not only be rescued from the trade, they need counselling and care. They need money and help. They need education and hope. They need love. It is awful to realize that many in this trade have never experienced *love*. Not once. They have only been used, exploited, abused and dehumanized. They need the love of Christ, and that love will only be mediated to them in and by the church. Like fighting against female genital mutilation, combatting sex slavery and the sexual exploitation of innocent children must be high on the list of priorities for the church's ethical engagement in the world today.

> ... combatting the sexual exploitation of children must be high on the list of priorities for the church today.

## CONCLUSION

Many books have been written on all the topics covered in this chapter. Christians need to recognize that the biblical ethic regarding sex is that sex is a good and pure gift, when placed in the right context. Sex is also much abused and distorted, which produces painful results. The world is awash in sexual immorality of all kinds, and Christians must stand against it. But this is not done merely by condemning—it is done by genuinely loving. In a world where sex is often made to be more important than relationship, and animal desire more important than the dignity of persons made in the image of God, the church of Christ

---

[3] We must recognize that in some places child prostitution is endorsed by religious practices. Although technically illegal, in India the *devadasi* are children who are married to temple gods. Worshipers engage in religious sexual acts with them. Christians *must* oppose all such practices, and where possible, insist that laws are enforced to protect the victims and prosecute the offenders.

ought to be a bright light for a better way. A person is not a sexual object. Sex is not the highest good or gift. When we truly love other people, we cannot dehumanize them or treat them selfishly. In our brokenness and sin, we are not whole in any area of our lives, including our sexuality. As a result, we will be merciful and gracious and compassionate to those who fall into sexual temptation and sin. We will point them to the forgiveness and healing and love that are found in Jesus. And, by God's grace, we will keep ourselves from sexual immorality, remain sexually pure and trust in Christ's redeeming blood. If anyone is in Christ, they are part of his new creation. Jesus is the one person who will never abuse us, never take advantage of us, never dehumanize us, never objectify us. He is love.

### REFLECTION QUESTIONS

1. What is your society's view of homosexuality? In what ways does it reflect the biblical view, and in what ways does it depart from Scripture?

2. Does your church have practical opportunities to help prostitutes or sex slaves (either in your own area or overseas)? If so, what are you doing to help?

3. Given the brokenness of our sexuality, how can the church bring healing to those who have committed sexual sins?

4. What kind of a biblical and theological case can you make for the goodness of sex in its proper context?

# 6

# *Abortion*

Although estimates vary, we can say conservatively that between 40 and 50 million abortions are performed throughout the world every year. For perspective, the estimates for the total number of deaths caused by the Second World War range between 50 and 80 million. World War II lasted for six years. This means that in a six-year period, approximately four to six times more abortion deaths occur than the total number of deaths caused by the bloodiest war in human history. Mathematically, in a six-year period, there are possibly 300 million abortions performed in our world. If the unborn are human beings, this represents an unimaginable loss of life. World War II ended and the killing stopped, but abortions continue at pace year after year. Matters of life and death are of the highest urgency and importance, so it is absolutely necessary for Christians to take abortion seriously. We need to pray, think and also act.

Not every person who has an abortion does so for the same reasons or with the same desires and motivations. Some governments have imposed limitations on family size, and women who become pregnant risk penalties for exceeding the limits. In North America, a major argument for abortion is that women simply have the freedom and autonomy to decide to have sexual intercourse, pursue a career or do anything else that they would like without having their plans altered by an unwanted pregnancy or baby. As a result, those who support abortion in North America call themselves "pro-choice" or say they "support a woman's right to choose" whether they want to keep the baby they are pregnant with. In developing countries, many women have abortions because they feel that they are not in an economic position to sustain a larger number of children. There are also many cases, of course, when abortion may be used to cover up adultery or illicit sexual relations. Shame, guilt and fear are transcultural motivators.

## A SIMPLE DEDUCTION

In order to formulate a basic position on abortion, two very simple questions need to be considered. First, is killing an innocent human being morally wrong? It will be assumed without argument that the basic answer to this question is, "Yes." Having granted this principle, we then consider the second factor: Are the unborn actually human beings? If they are, then taking their lives is an unjustifiable act of killing. As simple as this is, it is amazing how many arguments in favour of abortion fail to come to terms with this reality: If the unborn are human, ending their lives is morally wrong. Killing an innocent person because the mother wanted to have sex but didn't intend on becoming pregnant is not morally acceptable. Killing an innocent person because the family doesn't want an extra mouth to feed is not morally acceptable. Killing an innocent person because they are dependent on their mother's body and the mother finds that arrangement inconvenient is not morally acceptable. When Christians discuss the issue of abortion, they need to constantly bear in mind that the identity and nature of the unborn as *fully human* is absolutely crucial.

*If the unborn are human, ending their lives is morally wrong.*

## BIBLICAL DATA

Throughout its history, the church of Christ has taken a uniformly negative view about abortion, which was the same view that was universally shared in ancient Israel. The position in the traditional Judeo-Christian view is that abortion, simply put, is morally wrong. One could argue that in both the Old and New Testaments, in both Israel and the Christian church, abortion was actually almost unthinkable. In Israel, having children was considered an enormous blessing, and bearing children was prized by women in Israel's culture even more highly than it is in many parts of the world today. A woman's desire for children in ancient Israel was biological, instinctual, practical and also motivated by extremely strong social factors. Abortions were just not something that Israelite women were looking to procure in the ancient world. The only references to abortion that exist in the early church show a total rejection of its legitimacy. As a practice, it was completely condemned. Although some people have tried to make a case that the Bible permits abortion, their arguments strain credibility. Without question, the Bible does not support abortion—more so, its entire spirit is against it. The real issue is whether people will accept the Bible's authority. Both anti-abortion proponents and the vast majority of pro-abortion advocates recognize that the Bible is against the pro-abortion position. Since evangelical Christians acknowledge the authority of Scripture, they must oppose abortion.

> *Since evangelical Christians acknowledge the authority of Scripture, they must oppose abortion.*

### 1. Genesis 1:27-28

> So God created mankind in his own image; in the image of God he created them; male and female he created them. God blessed them and said to them, "Be fruitful and increase in number...."

Rabbinic commentators understood this passage to represent both a blessing and a command. Although there is nothing in the text that speaks directly of abortion—why would there be?—there is a creation principle that favours life. Much, much more can be said about the significance of this text for theological and biblical studies, but the first accent in Scripture is positive in regards to reproduction. This tone is perfectly in keeping with the remainder of the canon, where life and fruitfulness are seen as blessings. God blesses life and charges the human race with reproduction; abortion is antithetical to this principle.

## 2. Exodus 20:13

"You shall not murder." The Decalogue[1] begins with drawing the hearer's attention to the holy God who is Israel's Redeemer. These are God's words, not merely the words of Moses. Although the Decalogue is not the entirety of the Law, and although it is not exhaustive of God's moral will, Jews and Christians have recognized the special place that it occupies in God's unfolding revelation. This commandment is quite clear in its meaning—it forbids the taking of innocent lives. All human life—whether in the womb or outside of the womb—is protected by this command. In some Christian traditions, the Decalogue does not merely restrict negative treatment, it also enjoins positive behaviour. If that interpretation is followed, then the commandment not to murder is also commanding that positive care and protection be given to the innocent.

## 3. Exodus 21:22–25

> When men strive together and hit a pregnant woman, so that her children come out, but there is no harm, the one who hit her shall surely be fined, as the woman's husband shall impose on him, and he shall pay as the judges determine. But if there is harm, then you shall pay life for life, eye for eye, tooth for tooth, hand for hand, foot for foot, burn for burn, wound for wound, stripe for stripe (ESV).

This text is the most controversial passage in the canon when it comes to abortion. There are very difficult issues for both translation and interpretation. Some interpreters believe that the text is saying that if a pregnant woman is struck and she suffers a miscarriage, then the assailant suffers according to the principle of *lex talionis* (i.e., eye for eye, life for life) on the basis of what happened to the *child*. Other scholars, however, believe that the text is saying that if the child dies the assailant is fined, but if the woman dies then, and only then, is the

---

[1] The Ten Commandments. See Exodus 20:1–17 and Deuteronomy 5:6–21.

principle of *lex talionis* applied. Cases for either position can be made from the grammar of the Hebrew text. The view that a miscarriage occurs and that the loss of the child is not considered to be as serious as the loss of the mother is by far the majority position in the history of interpretation. This view is bolstered by the fact that a blunt trauma capable of inducing labour would almost certainly result in fetal death or catastrophic injury. It is difficult to maintain that this scenario envisions a pregnant woman being struck with sufficient force to induce labour, but that both the mother and the baby are entirely fine afterward.

Although we believe that a linguistic case can be made for applying *lex talionis* if either the baby or the mother are seriously harmed, for argument's sake we will take the position that the assailant is only fined if there is a miscarriage, but subjected to *lex talionis* if the mother is harmed. This interpretation places a value on the life of the mother that is higher than the value of the unborn child. Does such an interpretation mean that the unborn child is not considered a human being? Absolutely not! Exodus 21:18–36 deals with punishments for injuring others in a variety of circumstances. If a bull that was known to be dangerous gored a free person to death, the bull's owner was put to death for their wilful, criminal negligence. But if the bull gored a slave to death, the owner was merely fined. The fact that one case brought *lex talionis* and the other brought a fine does not mean that slaves were not considered human beings (in fact, Israel's law insisted that they were). Throughout the Pentateuch, there are also different values assigned to males and females, but there is no suggestion that males are *more human* than females.

Given this structure of valuation, it is illogical to say that just because the death of an unborn child was punished by a fine, but the death of the mother was punished with capital punishment, the baby was not considered a human being. The fact that the assailant was fined if the child was harmed indicates that its life was valued. The loss of fetal life was serious to both the mother and the father. The loss of

> **lex talionis**
> Latin, "law of retaliation." It refers to a form of punishment that corresponds to the offence committed (i.e., eye for eye, tooth for tooth).

the mother and wife was a greater loss to the husband and family, however, and it was punished more severely. This text may give some slight support for the position that abortion is permissible in the rare and extreme cases when the mother's life is in danger due to her pregnancy, but it in no way can be fairly used to suggest that the Bible supports a freely chosen act of abortion. The scenario depicted in this text is one of violence suffered and tragic loss—it is hardly a desired, sought-out and free choice.

### 4. Psalm 22:9-10

> Yet you are he who took me from the womb; you made me trust you at my mother's breasts. On you was I cast from my birth, and from my mother's womb you have been my God.

Psalm 22 begins with a cry of lament. David is calling out to God, feeling forsaken. (Psalm 22:1 is also cried out by Jesus on the cross.) As he suffers and is tormented, David recalls that before he was weaned, God's hand had been upon him. In fact, God watched over David even before David was born. God was intimately concerned with David in utero. David was a person whom God cared for while he was still in the womb. David's personhood, humanity and value were not conferred upon him only after he was born.

### 5. Psalm 51:5

"Surely I was sinful at birth, sinful from the time my mother conceived me." As David repents of his sin with Bathsheba, he looks back for a time when he may have been free from sin. The reality is that he has always been sinful, even from the very beginning of his life—which was at the moment of conception. There was nothing sinful in his parents' sexual act that conceived him, but from the moment of conception David was a person who inherited a sinful nature. One could argue that David is merely being poetic and rhetorical (it is a psalm, after all), but it seems that David really is saying that his entire personal existence has been in sin, and this personal existence started when he was conceived. Paul argues in Romans 5 that all people inherit a sin nature from Adam, and David asserts in this psalm that he was sinful at conception. Putting these two things together, it seems that the

Bible teaches: 1. All human beings receive a sin nature from Adam; 2. The sin nature exists from the time of conception. If all human beings inherit a sin nature, and if the timing of this inheritance is the moment of conception, then human life begins at conception.[2]

### 6. Psalm 139:13-16

> For you created my inmost being; you knit me together in my mother's womb. I praise you because I am fearfully and wonderfully made; your works are wonderful, I know that full well. My frame was not hidden from you when I was made in the secret place, when I was woven together in the depths of the earth. Your eyes saw my unformed body; all the days ordained for me were written in your book before one of them came to be.

These are the most popular verses quoted by those who oppose abortion from a biblical perspective. It is clear the psalmist is using poetic language which is not to be taken literally (the child is not literally being assembled in the core of the earth, of course). But it is very, very clear that the psalmist sees the sovereign God at work, forming and fashioning him before he was even born. God's work in forming children in the womb is wonderful. Even in the womb, the human body is fearfully and wonderfully made by the intentional and loving hands of God. There is a reason why these verses are so frequently

---

[2] In systematic theology, there are different positions taken in regard to when a human soul is created. "Soul creationism" is the position that God supernaturally creates every human soul *ex nihilo* and joins it to a body. "Traducianism" is the position that human souls are formed at the moment of conception through a merging of material from the parents' souls. In Traducianism, parents contribute physical genetics and also spirit "genetics." There are arguments for both of these views that we will bypass for now. Suffice it to say that a soul creationist may theoretically believe that God implants a soul at the moment of physical conception, or at some other time (such as implantation or quickening). Traducianists, however, believe that a soul and body are conceived at the exact same time. In this latter model, then, human life *must* begin at conception. In soul creationism, it *may* begin at conception or another time. Soul creationists, however, should take their view of the origin of full human life from the biblical data, and not from philosophical speculation. Soul creationists should affirm, therefore, that there is no reason to think that a human soul comes into existence at any time other than conception.

quoted to support the full sanctity of human life, even when it is in the womb. An honest reading of this text reveals an ethos that simply does not resonate with the position of those who support abortion.

### 7. Jeremiah 1:4–5

> The word of the LORD came to me, saying, "Before I formed you in the womb I knew you, before you were born I set you apart; I appointed you as a prophet to the nations."

Jeremiah was personally known by the Lord before he was born. In fact, he was known by God before he was even conceived. The person who was appointed to be a prophet to the nations was set apart by God before birth. There was continuity in God's mind between the person in the womb and the person who functioned as a prophet. When God says that he "knew" Jeremiah, he is using language that is loaded with deep relational intimacy. It is the word that is used for Adam knowing Eve (Genesis 4:1), and of God having an exclusive relationship with Israel out of all the nations on the earth.

> **When God says that he "knew" Jeremiah, he is using language that is loaded with deep relational intimacy.**

It could be argued that God only knows people in the womb who will be born, rather than every fetus in the womb whether they will be born or not, but this logical possibility does not overturn the clear fact that while the person of Jeremiah was in his mother's womb, he was known by God. This demonstrates that at least *some* infants in the womb are considered human by God. What evidence is there that some other infants are *not* considered fully human by God? Scripture doesn't reveal any such instances.

Unless we know for sure that the fetus is not a human being, terminating its life is irresponsible and unethical. In Canada, a hunter who shoots into the woods and kills a person is legally liable, since the law mandates that a hunter can only shoot when they *know for sure* what they are aiming at. If the hunter isn't sure what's behind the bush—perhaps it's a deer, a bear, a dog or a person—and they shoot, they are criminally responsible if they injure a person. This very obvious principle of responsibility should be applied in discussions of abortion.

If there is any doubt whatsoever, it is morally irresponsible to kill. It may also be criminal.

## 8. Luke 1:26–56 (especially verses 41–45)

> When Elizabeth heard Mary's greeting, the baby leaped in her womb, and Elizabeth was filled with the Holy Spirit. In a loud voice she exclaimed: "Blessed are you among women, and blessed is the child you will bear! But why am I so favored, that the mother of my Lord should come to me? As soon as the sound of your greeting reached my ears, the baby in my womb leaped for joy. Blessed is she who has believed that the Lord would fulfill his promises to her!"

There is no finer text in Scripture to teach the importance, value and personhood of the unborn than these verses in Luke. Elizabeth is in her sixth month of pregnancy, and the baby (Greek: *brephos*) leaps in her womb in response to the Spirit and the presence of the incarnate Son of God. Luke 1:15 makes it clear that Elizabeth's baby—John the Baptist— will be filled with the Spirit while he is still in the womb. Mary goes to find Elizabeth, and when they meet, Mary is in the early stages of her pregnancy. The unborn John the Baptist leaps in his mother's womb, because the Holy Spirit prompts him to testify to the presence of the Lord, who is in Mary's womb! John the Baptist's prophetic role was to prepare the way for the Lord, and he is blessed because he is the one who points out Jesus as the Lamb of God who takes away the sin of the world. Other prophets predicted the coming of the Messiah, but John the Baptist physically points to him and verbally identifies him. This prophetic function, however, is not discharged for the first time at the event of Jesus' baptism: it is discharged in this text, while both John the Baptist and Jesus are in the wombs of their mothers.

When Elizabeth says that the "baby" (*brephos*) leaped in her womb, she uses the same word that the angels use in Luke 2:12 and 16 when they announce to the shepherds that they will find the baby/*brephos* in the manger. John the Baptist is a *brephos* in the womb; Jesus is a *brephos* in the manger. We have biblical warrant, therefore, to speak of unborn *babies*. Neither John the Baptist nor Jesus began to live and to be human persons only after they were born. Jesus was specially conceived by the Holy Spirit in Mary's womb. He was the incarnate Lord of glory—fully God and fully man—from the moment of conception. If Elizabeth had aborted her baby, she would have killed Messiah's prophet. If Mary had aborted Jesus, she would have killed the Messiah, the Son of God incarnate. Personal identity, life and the Spirit's work are evident in this narrative long before either baby was born. Jesus would have been at the earliest stages of fetal development when the Spirit moved in the unborn John to leap in recognition of the Messiah.

## ARGUMENTS FOR AND AGAINST ABORTION

From a biblical perspective, it should be very clear that abortion violates the will of God. Not everyone, however, reasons from—or even respects—the biblical worldview. In every particular ethical issue, there are broader and more basic foundational blocks that we need to consider. (We discussed some of these issues in Chapter 2.) It is always worth bearing in mind that not every worldview can make sense of morality, ethics and the concept of human rights. When someone insists that a woman has a right to choose an abortion, it may be worth asking *where* that right comes from, or why the woman's preference demands moral acceptance. In other words, we often forget that ethical discussions have to be located in proper moral frameworks: not every structure can support moral rights. We need to be wise in what we say, when we say it and how we say it, but sometimes we need to address moral foundations before we can address specific ethical issues. This is especially the case when a person's worldview generates internal contradictions. As previously noted, secular humanism is not able to provide a sustainable and cogent account of the relationship between the nature of human beings and their alleged moral rights. As a result, many of the arguments that abortion advocates use will not only break down in terms of the actual argument, they will also fail to be logically coherent inside of the advocate's worldview.

### 1. *Prudential arguments*

Some pro-life advocates believe that a convincing case can be made against abortion on the basis of the harmful effects that women can suffer from having one. For example, there are a number of physical complications for the woman that can arise, and future fertility can be negatively affected. Many pro-life proponents also claim that a large number of women experience "post-abortion trauma," and other psychological maladies that can be marked by nightmares, extreme guilt, disruption of trust between partners, and even an increase in attempted suicide. These claims are dismissed as fanciful by many of those who support abortion.

One of the inherent difficulties with all prudential argumentation is that different people can define "harm" in different ways. For some people, little could be more harmful to women than having an unwanted pregnancy that disrupts their economic independence or career. It is possible to maintain that even if an abortion procedure results in a net harm for a particular woman, the cost may be worthwhile if abortion serves to emancipate women in general. On this utilitarian calculation, the harm some women suffer might be offset by the good that the majority of women gain. It is also the case that abortion procedures could become safer and safer, thus reducing the possible physical complications. Better counselling services and support could also help women recover more quickly and with less trauma. So although a good case can be made that abortion can in fact produce physical, emotional and psychological harm in some women, this style of prudential argumentation can yield different conclusions depending on how terms are defined, what one's priorities are and how theory is put into practice.

The same type of considerations exist when it comes to arguing against abortion on the basis of the violence of the procedure itself, rather than focusing on the result of the procedure (which is the death of the child). It would be extremely easy to present very graphic descriptions of what various abortion procedures entail, but we will be constrained. Suffice it to say that abortion procedures can involve the literal dismemberment of the baby, burning the baby to death in chemical solutions and ripping the baby's body out of the womb. It is a scientific and medical fact that these procedures are often practiced after the time when the fetus has developed the ability to feel pain. If too much emphasis is placed on the brutality of the procedures, however, the most important point may be lost. Even if the death of the unborn child could occur painlessly, this would not make it a morally acceptable act.

**Even if the death of the unborn child could occur painlessly, this would not make it a morally acceptable act.**

## 2. Scientific data

The field of embryology has revealed many fascinating facts about the development of human life in the womb. We can watch and track the way the fetus develops during the gestation period. Scientifically, we know that at the moment of conception the entire genetic code of the baby is present. There is never a time when the conceptus lacks even a little bit of its full genetic material. Since it is the genetic material that drives the organism's development, it is impossible to say that a living organism that has its full genetic code is somehow less than, or something different from, what its genetics say that it is. Whatever the stage of development—zygote, embryo, fetus—genetically speaking, it can be nothing other than human. This means that whatever the developing baby looks like, it looks like *what a human looks like* at that particular stage of development. There is a time when a human life is only eight cells. There is a time when a human life does not yet have a beating heart. There is a time when a human life does not have a fully formed and functioning brain. But the fact that these features will develop over time does not mean that the organism is not human until these features exist. In fact, the only reason that a human brain will develop is because the baby is genetically a human being.

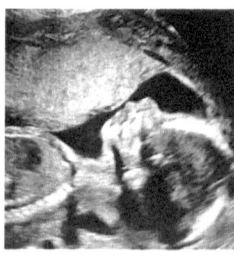

An ultrasound of Grace Penhearow at 18-weeks reveals how far physical development has progressed since conception.

In North America, some proponents of abortion have argued that women should not be shown ultrasound pictures of their unborn child before they make a decision about abortion. The reason for this, interestingly enough, is that when women see the humanity of their child, hear its heartbeat, see its outline, etc., they often cannot bring themselves to terminate its life. They intuitively recognize that the fruit of their womb is not a blob of inanimate tissue—it is a living human being. It is disturbing that some who call themselves "pro-choice" want to limit a mother's access to pertinent information which might influence her decision making.

There are some people who believe that a woman should have the right to an abortion only during their first trimester of pregnancy, due to the relatively undeveloped physical and mental state of the fetus. Yet, as has been noted, at the very moment of conception the child is fully human on the basis of their genetics. In the first month of pregnancy, this human being implants into the mother's uterus and begins to experience critical cellular divisions and organic growth. In the second month the baby has brain waves, a heart that begins to beat, facial features, and is sensitive to environmental stimuli. In the third month the baby can suck their thumb; they can also feel pain. Their form is recognizably that of a human baby. These few points are not even remotely exhaustive of the incredible development that occurs in the first trimester of pregnancy. It is absolutely certain, however, that there is no time or stage when the child moves from being "nonhuman" to "human." Scientifically, from the moment of conception on, there is a full human person who experiences growth and organic human development.

> **Scientifically, from the moment of conception on, there is a full human person who experiences growth and organic human development.**

### 3. Quality of life

The term "quality of life" is a very important term in discussions of both abortion and euthanasia. A major division occurs between those who see "quality of life" as outweighing the inherent "sanctity of life." Beyond this, it is next to impossible to get any clear definition of what "quality of life" actually means. Qualitative considerations are invariably preferential, and therefore they cannot be forced on anyone else. Since quality of life seems to evade definition, case studies and hypothetical examples abound. Normally the idea is that if the unborn child will be born with a terminal condition (like Tay Sach's disease), it may be more merciful to abort them so that they don't suffer for a short period of time after they're born. Others have suggested that abortions should be performed on fetuses that will be born with either severe physical or mental disabilities. Famed atheist Richard Dawkins went on public record as saying that it would be immoral *not* to abort a fetus with Down Syndrome if the option was available. The outcry against his statement was, thankfully, enormous.[3]

---

[3] One wonders, of course, where Dawkins gets his high moral code from. As a materialist, it is amazing that he not only accepts the existence of genuine morality,

Proponents of the quality of life argument should be somewhat nervous that their understanding of what quality of life means will be unavoidably subjective and relativistic. Sanctity of life proponents, on the other hand, have an objective ground on which to stand. In this view, every human being is intrinsically worthy of life, and no innocent human being should be put to death. Quality of life, on the other hand, is in the eye of the evaluator. What one person sees as an unacceptable quality of life may be quite acceptable to someone else. It has been pointed out that there are many groups that work with people with Down Syndrome, and not a single one of these groups endorses killing fetuses—or toddlers, or teenagers, or adults—on the basis of their having this condition. If aborting an unborn baby with Down Syndrome is justifiable on the basis of quality of life considerations, why can't we kill Down Syndrome children three months after they've been born? If it's about quality of life, it is difficult to maintain that the prospects for quality of life improve simply because the child has been born into the world. As with so many arguments for abortion, the principle that is supposed to justify abortion also justifies infanticide.

### 4. Babies are not persons

Some philosophers maintain that being a full human person requires self-consciousness. Since unborn children do not have a fully formed self-conscious awareness and understanding of their own existence, these thinkers argue that they should not be considered full human persons. This metaphysical distinction between "human" and "person," however, is obviously not taught in Scripture. The entire construct is philosophical and subjective. Who says that self-consciousness is a necessary condition for full human personhood? Again, if a human life is present at the moment of conception, we are within our rights to say that it logically follows that self-consciousness is not absolutely necessary for full human personhood. There are stages of human development where self-consciousness has not yet been formed, but it is still a full human being who is developing new faculties.

---

but apparently he knows that failing to abort a child with Down Syndrome is an immoral act. Not only was his suggestion insensitive and ignorant, it was also ludicrous in terms of his own worldview. These are the kinds of times when it is worthwhile to push back on someone, exposing the bankruptcy of their moral beliefs in relation to their worldview foundations.

It is also the case, of course, that there are times when self-consciousness can be lost. For instance, degenerative brain diseases can destroy self-consciousness. Traumatic injury can do so as well. It could even be argued that we are not self-conscious whenever we are unconscious, which might include when we're sleeping! Yet somehow it doesn't seem morally acceptable to kill people in their sleep. In response, some say that killing a person in their sleep is morally wrong because it robs that person of their potential future experiences: if they were awake (i.e., self-conscious) they would object to being killed. Precisely. Why would we think that the fetus would choose to die and be robbed of their future, rather than think that—like us—the fetus would choose to live if they were given the option? When we reason from our own experience, we know that we are glad that we were allowed to develop into self-conscious beings.

This issue also raises the question of infanticide. Some philosophers (like Peter Singer and Michael Tooley) have argued that infanticide is morally acceptable. They are honest that the arguments used to support abortion logically apply to the killing of newborn infants as well. Since a baby born one month ago is not fully self-conscious, *if* the self-consciousness argument is valid, it should be permissible and legal to kill the baby. We should also have a sliding scale for the value of different human persons. How self-conscious do you need to be in order to qualify as a fully human person? A three-year-old is more self-conscious than a one-year-old. A twenty-year-old woman has deeper self-consciousness than an eight-year-old girl. Perhaps a person in the advanced stages of Alzheimer's disease loses their personhood as their mind deteriorates, and eventually reaches a stage of such low self-consciousness that their family can kill them. Once the concept of human personhood revolves around a philosopher's subjective evaluation, the floodgates are opened for killing all who do not measure up to their relativistic standard. History, in fact, has known regimes that

> **Once the concept of human personhood revolves around a philosopher's subjective evaluation, the floodgates are opened for killing all who do not measure up to their relativistic standard.**

have murdered innocent millions on the basis of similar lines of reasoning.

### 5. Unwanted pregnancies lead to child abuse

This emotive argument is quite common. The idea is that a woman should be able to abort her child if she doesn't want it, because unwanted children may be abused or neglected. Both the conclusion and the premise of this argument are scarcely proven to be true. Is it really necessarily the case that a mother who didn't want to be pregnant will not love her child when it is born? Furthermore, is it really being contended that the mother and father who don't want the child are so immoral and wicked that they will abuse their child after birth? Not wanting to bear a child is one thing; abusing that child is quite another. There are laws against child abuse, and they need to be enforced. Surely no judge would excuse a parent if they abused their child but pleaded that they only did so because they didn't want the child in the first place. This argument also misses the entire point of abortion: If the unborn child is aborted, they have suffered the very worst form of abuse possible—they have been violently killed. If a parent knows they cannot properly care for the child, there are still numerous options for that child to grow up in a loving environment elsewhere.

### 6. Women need access to safe abortions

Those who maintain that a woman should have the right to a safe and legal abortion often assert that many women will die obtaining illegal and dangerous abortions if they are not allowed to have one legally. There is some emotional force to this argument, and it is one that is heard very frequently. As popular as it is, however, it is extremely poor. **The truth is that if abortion were illegal, far fewer women would have one.** Whenever a country legalizes abortion, the number of abortions performed in that country rise exponentially. It is simply wrong to assert that all of the abortions performed every year would continue to be performed in unsafe locations. Even if women wanted an abortion, they may not risk the legal penalties (just as many people would like to walk out of a store without paying for items, but fear of prosecution keeps them from doing so). Pragmatically, pregnant women may also be smart enough not to have a back-alley abortion, since they would know the possible outcomes and risks.

What is far more important, however, is that this argument rests on two massively mistaken notions. First, it completely begs the question of whether there is such a thing as a *safe* abortion. Every time there is an abortion, the result is the death of an innocent human being. A "safe abortion" is an abuse of language. Second, the argument assumes that the government should make it safe for people to do certain immoral or illegal things. If abortion kills an innocent human being, does the state really have a moral responsibility to ensure that this killing is done in a way that doesn't harm the killer? This is strong language, but it needs to be used. Imagine that there were 50 million legal murders per year in the world, and some people thought such murders were outrageous and should be made illegal. If a pro-murder advocate insisted that we had a moral responsibility to make it safe for murderers to kill their victims, it is doubtful that we'd be convinced. Probably most people would be more concerned for the innocent victims than for the murderers. If somebody chose to engage in an illegal act of murder and was hurt in the process, it would seem that the responsibility and guilt would be upon their own head. Concern with the danger of "back-alley murders" for the sake of the murderers, combined with the proposal that the state should provide funded and safe murder clinics to ensure only the victim died—this is simply a bizarre position to take. The principle is deeply, deeply flawed, and assumes the unborn are not innocent persons.

> **Every time there is an abortion, the result is the death of an innocent human being.**

### 7. Women should have control over their own bodies

At one level, it is difficult to disagree with the principle that women should have control over their own bodies. Men ought to have control over their own bodies, too. But having control over our body is limited by the law when it comes to how our body can affect other people. We have the legal right to swing our arms through the air, but not when somebody's face is occupying the space where we are aiming our fist. In many countries, it is illegal to commit suicide, and those who attempt to do so are forcibly restrained. Those who want to severely cut and harm themselves are also restrained. We simply do not have unlimited rights to do whatever we want with our bodies. We are also responsible for the consequences of our bodily actions. If a woman has consensual sexual intercourse and becomes pregnant, she needs to accept responsibility for her actions. So does her partner, of course. Now, it is easier for the male partner to abdicate responsibility than for the woman (although that doesn't mean that the male is not morally liable for his failure to act properly). The woman who becomes pregnant will have nine months of carrying the baby, and will then either have to provide the necessary care or give the baby to someone who can meet its needs. This will affect the woman's body, and possibly her economic position, education plans, social status and various other things. But, it is essential to remember that the woman's economic status does not justify killing an innocent person. A woman who engages in sexual activity runs the risk of becoming pregnant, even if precautions are taken. If she conceives, she and her partner are fully responsible. Killing an innocent person is not permissible just because she doesn't want to be pregnant.

> **We simply do not have unlimited rights to do whatever we want with our bodies.**

Once again, if this principle justifies abortion, it also justifies infanticide. A baby is completely dependent for months. What if a mother decides that feeding and caring for the baby is too much of a burden? What if the mother doesn't want to use her legs to walk to the crib, or use her breasts to nurse, or use her hands to change diapers? If a woman has autonomous control over her own body, then she shouldn't have to use her body to care for an infant that she doesn't want. It does not make a morally relevant difference if the baby is dependent on her in the womb, or if the baby is dependent on her outside of the womb.

> **Changing physical location does not change a being's essence or nature.**

This leads to a very important point: Changing physical location does not change a being's essence or nature. What relevant difference does it make to the child's nature if it is in a womb, or if it is outside of the womb? As already mentioned, some abortion advocates are logical enough to acknowledge that their position also supports infanticide. Failing to observe this point has led to some incredibly confused and contradictory situations in certain hospitals. In some hospitals, doctors are called upon to kill a nine-month-old unborn baby, and then to turn around and work to save the life of a prematurely born infant that is *months younger* than the unborn child whose life the medical staff just ended. A procedure called "partial-birth abortion" involves killing the baby as it exits the mother's birth canal, just before a successful natural delivery. Is it really feasible to maintain that the child that is halfway out of the mother's body is a non-person who can be killed, but if its other half comes out, it is transformed into a fully human person who needs to be protected by law? How can a matter of inches be the difference between having full human rights and being fully disposable? How can a matter of inches make the difference between a murder and a legal killing? A being's nature does not change on the basis of physical location. A fish is a fish whether it is in the water or in a boat; a human is a human whether in the womb or outside of the womb.

> **How can a matter of inches make the difference between a murder and a legal killing?**

### 8. Abortion in the case of rape

There are Christians who disagree with virtually all elective abortion procedures, but who believe that an abortion may be justifiable if the child was conceived by an act of rape. We must not diminish the horror and trauma of being the victim of rape. Those who have been raped should receive the maximum level of support possible. They need both love and justice, and Christians should lead the way. It is agonizing and heartbreaking to consider the shame that a woman may feel if she conceives a child when she is violated, even though she is entirely innocent and blameless. Throughout the pregnancy, people would ask her about the father, and make assumptions about her circumstances. Pregnancy cannot be hidden, and it may be a constant reminder of the trauma of the rape. We do not have any adequate words for this situation. It is shattering.

With as much sensitivity as we can muster, however, we need to examine whether a conception caused by rape is justification for having an abortion. For a variety of reasons, conception only occurs in a very small percentage of rapes. No more than one per cent of the total number of abortions that take place were performed because the woman was raped or the victim of incest. This means that 99 per cent of abortions have nothing to do with rape; they are procured for other reasons. Speaking as gently as possible, the *circumstances* of the conception are not relevant to the nature of the being that is conceived. If human life begins at conception, then the circumstances of conception do not alter that fact. An immoral act can lead to conception, but this doesn't change or alter the nature of the person who was conceived. Once a human being exists, they should be protected from unjustified killing. They have a right to life. Aborting the child punishes the child with death for the act of their biological father. It is the rapist, not the child, who should be punished. Two wrongs do not make a right, and ending the life of the innocent child does not become morally acceptable just because the rapist perpetrated a horrific and violent crime. If the mother finds the presence of the child too difficult and painful, the child can be given up for adoption.[4]

## 9. Abortion to save the life of the mother

This is the most difficult case of all for those who uphold the sanctity of life. There are some cases (like ectopic pregnancies), where the baby is guaranteed to die no matter what is done to save it. Sometimes the baby is also a threat to the life of the mother. Is it permissible to have an abortion if such an act is necessary to save the life of the mother? Many pro-life advocates believe that it is, and this is the only circumstance in which they would allow an abortion to be performed. It is important to note, however, that even in permitting abortion in this circumstance, the procedure would still be done with a sense of grief. The loss of any life is tragic, even if it is necessary.

---

[4] If the mother finds that the child in her womb is too painful as a reminder of the rape, abortion advocates say that she should have the right to terminate its life. But the same principle could also be extended after the child is born. What if the mother originally decides she wants to keep the baby, but then two months after it is born she finds that it reminds her of the rape she experienced? If she can kill the baby when it is in her womb for that reason, why doesn't the same principle apply after the birth? Again, the issue comes down to the nature of the being: If the child is a human being, its life should be protected whether inside or outside of the womb.

As we saw in Exodus 21, the Bible may give just the slightest bit of support to the view that the life of the mother outweighs the life of the child. Given the ambiguities in the text, however, and the fact the law assigns different values to males and females, this evidence is only of the tiniest bit of weight. What is required is an attempt to uphold the sanctity of life of both the mother and the unborn child. If the baby is going to die, however, and it is possible to save the life of the mother, then an abortion may be justified on the basis of the principle of double effect. This principle of medical ethics states that a procedure can be performed if the goal is positive, if it is necessary and if the positive outcome outweighs the negative consequences. What is crucial for this principle is that the negative consequences are not desired or intended; nevertheless, they are sadly unavoidable. When it comes to the present issue, the idea in applying this principle is that the doctor is not trying to *kill the child*, they are trying to *save the life of the mother*. If a procedure to save the life of the mother results in the death of the child—as grievous and unwanted as such an outcome is—it might be justified on the basis of the fact that the death of the child was necessary for saving the mother's life. If both mother and child are going to die, but the mother may survive if her baby is aborted, this may be the only circumstance in which an abortion procedure is permissible. Of course, the number of abortions that are performed for this reason are an exceptionally low percentage, even though they are personally tragic for the mothers who find themselves in this position.

### 10. Who will care for the baby?

Christians must not only oppose abortion, they must work to provide love and support for women who are struggling in their pregnancies and in raising their children. No woman should feel hopeless or out of options if she becomes pregnant, whether she is married or not. Churches and individual believers should ensure that their hearts and hands are open to help the vulnerable. In North America, there are many organizations that were created and exist to help women receive counselling

**Christians... must work to provide love and support for women who are struggling in their pregnancies and in raising their children.**

and practical support throughout their pregnancies and even long after they've given birth. Love is required, but so is intelligent organization and infrastructures of care. Those who believe in the dignity and sanctity of human life should demonstrate it, not by articulating it in theory only, but by living it out. We are not called to love humanity *generally*; we are called to love *specific people particularly*. Our "neighbours" may be pregnant and needing help, and we need to help them, for their sake and for the sake of their child.

**Adopting infants and children reflects the heart of God.**

Instead of abortion, a beautiful and noble option is adoption. Adoption is one of the most wonderful of all the theological blessings that are ours in Christ. God the Father adopts us as his children, making us his heirs (Ephesians 1:5). Adopting infants and children reflects the heart of God. If women did not have abortions, many of them would keep their babies; but many others might give them up to be adopted. The church of Christ needs to be willing to share the adopting love of God that she herself has experienced. Until that day, we need to do all we can to provide love, care, counsel, support and practical resources for both women and their children. This will include sharing with them the forgiving and cleansing power of Christ's blood.

Abortion is a great evil, but it is not unforgiveable. Churches need to ensure that women know that the gospel of Christ is greater than all their sin, no matter what it is.[5] In Jesus, forgiveness can be found for abortion.

---

[5] It needs to be acknowledged that men can also experience guilt and shame over abortion. Men may be responsible for pressuring their partners to have an abortion, and in some cases they may finance the procedure. But even if the man opposes the abortion, he may feel guilty for not being able to stop it from taking place. He is, of course, responsible for his part in bringing about the conception. The termination of the life of his child—whether he desired it or not—may weigh heavily upon him as time goes by. It is only in Jesus Christ that forgiveness and healing can ultimately take place for either men or women, and there is grace for all who seek it.

I, Steve, remember being in the slums of a major city in Africa, seeing cheap paper advertisements plastered on poles and walls that advertised abortion services. The handwritten phone number on the bottom of the notices gave an indication of the quality of the service being offered. That there were so many advertisements (from competing groups) showed how popular the services were. What's the solution? Legalize the killing? Ensure that innocent, unborn humans die in a sanitized environment? No. The solution is to love and serve and give and work and pray. Abortion is a great moral evil—the death toll is unspeakable—but it is not taking place outside of wider social structures and human relationships. We need changes in both the hearts of individuals and in the structures of society. Only when the love of Christ is more real and vibrant to people than their hopelessness and fear; only when the sanctity of someone's life means more to us than our education, money, comfort or pleasure; only when we value others more highly than ourselves; then, and only then, will abortion end.

### REFLECTION QUESTIONS

1. What do you think the strongest argument is in support of abortion?

2. What do you think the strongest argument is against abortion?

3. Can you make an argument against abortion from an atheistic worldview?

4. What can you and your church do to stop abortion? What can you do to help pregnant women? What about children, both the born and the unborn?

5. What other biblical passages relate to this issue?

# 7

# *The end of life*

At certain times in Western history, it used to be said that Christians knew how to die well. And, in the main, that saying was true. It did not mean that every Christian was always unafraid of death. Even if unafraid of death itself, believers were certainly often afraid of the pain that might accompany the dying process. Someone might also doubt that they were really trusting Christ sufficiently for salvation. But, adjusting for all of that, Christians knew how to die well in comparison with others. They did not die with despair or angst, or with foolhardy bravado. They did not have to pretend that they didn't care about death, but neither did they have to rebel against their appointed end. Believers knew that they were in Christ, that death was the last enemy and that it was already defeated in Christ's death and resurrection (1 Corinthians 15:55–57). Death was the last remaining effect of the curse that they would ever experience. Death was the gate that marked the end of their earthly path, and in going through that gate, they entered their Master's estate. They finally arrived at their journey's end. They were finally home.

Knowing how to die well is what allows someone to live well. It is also necessary for making proper ethical judgements about the sanctity of life and medical care. Medical technology is very unevenly distributed around the world. Whereas some people can be kept alive almost indefinitely by machines that cost millions of dollars, others have no access to even primitive hospitals. There are medical options available to the rich in some nations that the poor in other places cannot imagine. Yet, not every medical advancement is an unadulterated blessing. Affluent medical health-care systems have the potential to do much good, but they can also produce much harm. They can make life increasingly artificial by reducing a person to the sum of their biological functions, rather than caring for the patient as a holistic person. They can also remove people from surroundings of love and care, replacing warmth and familiarity with the sterility of a hospital ward managed by overworked and under-appreciated staff. Patients can be reduced to units of economic gain for the private hospital, or units of economic loss for the public hospital. Whenever patients are viewed in terms of assets or liabilities, the medical environment is compromised. At their best, medical structures are compassionate and humane, where the health-care providers and the patients (and patients' loved ones) are all accorded love and respect. In many places, however, this ideal is not being experienced.

What word should Christians speak concerning death and dying? How can churches train their people to die well? When can Christians embrace medical technology, and when must we pull back? Whether rich or poor, are we countercultural when it comes to how we navigate the end of life? We ought to be. Our priorities, values, perspective and hope are quite different from the world around us. Since this is the case, our ethical commitments in matters of life and death will be different, too. We must avoid an unexamined acceptance of the norms of our surrounding cultures. Given our calling in Christ, we are to live our entire lives in the sphere of the sacred—and we are to die as holy people who are reserved for a holy God.

## MEDICINE, LIFE AND DEATH

When is someone actually dead? This may seem like a strange question to ask—the kind of question that only a person who is out of touch with the natural order of life can possibly raise. In history, death is as universal as birth. There is a moment when a person is alive (perhaps just barely and weakly), but then they breathe no more and are gone. There will be no more conscious movement, or light in the eyes, or response to stimuli. Human beings living close to the earth—farming, hunting, nurturing the sick in villages rather than hospitals—know the difference between life and death. Increasingly in the Western world, however, people are far removed from death. Medical intervention has the potential to blur the line between living and dying. Machines can keep hearts beating, lungs pumping and kidneys functioning. Through tubes and injections, nutrients and water can be put directly into the stomach and bloodstream. But this possibility has caused some people to wonder if a person is truly alive if they are permanently unconscious and non-responsive, while man-made machines sustain every function of their vital organs.

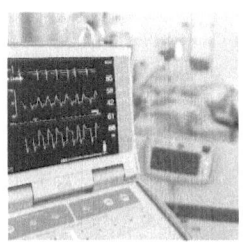

Modern medicine, especially in the Western world, can make ethical decisions surrounding end of life very complex and often challenging for families to navigate.

In the absence of advanced medical technology and infrastructure, such a scenario and such questions are irrelevant in a practical sense. Still, there are others who have stood in hospital rooms, trying to determine if their most dearly cherished loved ones who are permanently and unconsciously hooked into machines and lights and buzzers are really alive, or—in an important sense—basically dead already. One of the difficulties that attends such a scenario is that the obvious and natural signs of death are absent, and the machines keep these signs from occurring. For example, in a natural and common sense environment, when a person stops breathing and there is no pulse, they are pronounced dead.[1] Throughout human history, these were always the signs that were looked for, and they were definitive. Today, however, a person

---

[1] The ability to scan brains for activity has added a new layer to the discussion. Most ethicists today accept that when there is no brain activity (i.e., when a person is in a state of brain death), then the person is dead. But throughout history and in the vast majority of cases today, individuals are not hooked up to brain scan equipment. Lack of pulse and cessation of breathing are still the main indicators death has occurred, even in countries with advanced medical technology. We need to recapture the reality that death is not primarily a legal or philosophical question: It is pre-philosophical, a matter of human nature. Where nature and humanity have been lost sight of, we need to retrace our steps until we find them again. Technology, including medical technology, must be our servant, not our master.

can be in an irreversible coma, and they may be completely dependent on machines—never even knowing that they are alive. But is such a state one of *living*, or is it one of artificially prolonging *dying*? Answering this question is critical for a proper ethic of life and death.

## MEDICAL CARE: POSITIVE AND NEGATIVE

Countries that are privileged to have excellent medical care are truly blessed. Jesus' life and ministry demonstrated that God is on the side of compassion and healing. There is no theological reason to turn our backs on the provisions of medicine. All healing is ultimately from God, and he has healed miraculously, but God also heals through granting success to ordinary means. He gives doctors and nurses knowledge and wisdom, and it is right that they use their gifts to bring healing and wholeness wherever they can. Sickness, disease and death can and should be combatted. Quality medical care that restores health and either stops or slows down the ravages of illness is something for which to thank God.

Those who have access to good medications, procedures, clinics and hospitals, know that they would not want to be without them. Life expectancy in countries with excellent medical care is much higher than in places without it. Surgical procedures can dramatically improve quality of life, as can prescription pills. In fact, surgeries and medications can provide some people with decades of "extra" time on earth. Large numbers of people in the developed world have survived heart attacks, cancer, catastrophic traumas and all kinds of other potentially terminal conditions, because they had access to advanced health-care. Premature babies have survived and flourished. Mothers have lived through childbirth experiences that would have resulted in death in other times or places. Grandparents have had their lives saved through heroic medical intervention, allowing them to see their grandchildren grow up (and perhaps even see the birth of their great-grandchildren). Where such medical care exists, it is welcome and desirable. Medical care that sustains and improves life is morally good, and we are to be thankful for it.

*Medical care that sustains and improves life is morally good.*

The goodness of health-care is non-controversial—there are not too many people who are advocating for a reduction in its quality! But

*The end of life* 163

there does come a time when medical technology and procedures reach a point where a person's life is not being extended as much as death is merely being delayed. There are human ways of living in the world; nature can be a guide, albeit not an infallible one because we live in a fallen world. However, since nature is creation, we must learn to live in harmony with it. Just because a heart can be kept pumping by an artificial machine does not mean that it *must* be or even *should* be. The mere fact that we *can* do something is not a sufficient justification for doing it.

Recognizing, then, that advanced medical care is, in principle, morally good, we also need to recognize that its existence and use can create ethical dilemmas. When are people allowed to die? Should intensive, high-risk, painful and expensive procedures be performed on the terminally ill who are very old, and who will have very limited time left on earth even if the procedure is successful? We know that certain drugs that fight cancer can produce extremely harsh effects in the body. If someone has terminal cancer and is estimated to have three months to live unless they seek treatment, must they take these very harsh drugs in order to live six months instead of three? The drugs may slow down the cancer's growth, allowing the patient a few more months of life. But is the person really better off taking this treatment? Are six months of suffering, pain and perhaps mental fogginess, better than three months of somewhat normal yet decreasing function and relative clarity?

There are also many cases where people are kept alive by being connected to machines—machines that may circulate their blood, feed them and breathe for them—while they are permanently and irreversibly unconscious. If someone's heart cannot beat on its own, and their lungs cannot breathe naturally, and they cannot eat or drink on their own power, must they be plugged into a machine that artificially does those things for them? Even if it is ethically *permissible* to plug in, is it ethically *mandatory*? What is the purpose or goal? Is it okay to decide not to take certain medical treatments and simply to die instead? Although people should be able to pursue medical treatment if they want to, there is no moral imperative to pursue every possible medical treatment available. Life is good, but in a fallen world we need

**Is it okay to decide not to take certain medical treatments and simply to die instead?**

to take into account the ravages of sin and the inevitability of death. Only in Christ is death defeated; until he returns, everyone who is born will succumb to it. This means that our earthly lifespan is bounded, and we need to learn how to both resist death and yield to it. There is "a time to be born and a time to die" (Ecclesiastes 3:2).

## EUTHANASIA

The previous considerations lead to the question of euthanasia (a term which means "good death"). In many countries where advanced medical technology exists, death can sometimes be put off for long periods of time. But when can a person just decide that they'd like to die? When is life no longer worth preserving at all costs? These are questions that involve the notion of "quality of life." We measure the goodness of life not merely on the basis of longevity (i.e., quantity of time), but also on the quality of time. If given the choice, many people would rather die than spend months or years in severely debilitated conditions. When people are diagnosed with terminal and irreversible conditions that lead to extraordinary physical and mental breakdown and pain, many would prefer to die quickly and painlessly. Proponents of euthanasia often say that they are in favour of "death with dignity." This means that a person should be able to choose the moment of their death, before they deteriorate too far into severe mental anguish and physical pain. Once quality of life reaches a certain low threshold, and there is no prospect for recovery (and only the expectation of greater pain and further degeneration), those who support "death with dignity" believe it is ethically permissible for a person to choose to die.

> *...those who advocate for "death with dignity" tend to argue for a type of euthanasia called active euthanasia.*

What is highly controversial, however, is that those who advocate for "death with dignity" tend to argue for a type of euthanasia called *active* euthanasia. Active euthanasia is to be distinguished from *passive* euthanasia (although it may be difficult to fully separate them in hard cases). In active euthanasia, an individual performs an act that is the immediate cause of death. If active euthanasia is administered by a physician, nurse or some other party, then they are the agents whose actions are the immediate cause of the person's death. For example, a doctor might give a terminally ill patient an injection that puts them to sleep and then stops their heart from beating. Without this injection the person would continue to live, so their death is actively brought about by the injection. A doctor might also give a patient a pill to swallow that will have the same effect, but the patient is the one who actively takes it and swallows it, bringing on their own death.

Passive euthanasia can take different forms, some of which may seem more ethically acceptable than others. As so often, it partly depends on how terms are defined. Whereas active euthanasia involves administering something that brings about death, passive euthanasia involves withholding things that are necessary to keep the patient alive. For example, if someone requires a feeding tube because they can't swallow, removing the feeding tube ensures that they will no longer receive nourishment, and they will eventually die from starvation. Food is natural and necessary for life, so the removal of the tube that's required for the input of nutrients passively brings about death. Less controversially, if a person has been in a coma for a month with no prospect of recovery, and their bodily functions are only continuing because they are hooked up to machines, passive euthanasia would involve unplugging the person from the machines. In this latter case, some would argue that this isn't even passive euthanasia; it is simply a regular death. Others maintain that because medical care which had been applied is now removed, this withholding of the previously administered support qualifies as passive euthanasia.

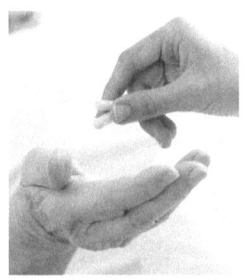

**Active euthanasia brings about death in an unnatural, pre-emptive and artificial fashion.**

A major distinction between active and passive euthanasia is found in the phrase, "let nature take its course." If someone can't survive apart from being permanently and unconsciously hooked up to medical machines while they lie in a hospital bed, then their life is being preserved by purely *un*natural means. There is nothing natural about the way that their life is being extended. Unplugging them from the machines lets "nature take its course." Active euthanasia does *not* let nature take its course. In fact, active euthanasia brings about death in an unnatural, pre-emptive and artificial fashion. God has given us permission to fight against illness, pain and disease, but he has not given us permission to actively put the sick to death. Since it is natural for living entities to strive to live, we can fight along the grain of nature to preserve life—we must not fight against the grain of nature to bring death. In the former case we are trying to *restore* harmony with natural health, wholeness and peace. In the latter, we are actively aiding the effects of the curse. Active euthanasia is the killing of an innocent person, and thus is by definition either suicide (if self-administered) or murder (if performed by someone other than the patient). Active euthanasia is therefore both unnatural and immoral.

> **Active euthanasia is the killing of an innocent person, and thus is by definition either suicide or murder.**

This judgement about the moral status of active euthanasia must be balanced, however, with sympathy. Unfortunately, in some cases, it is possible that family members may simply wish to see an elderly and infirm person die so that they can inherit their estate. It is possible that doctors will encourage the sick to die so that they don't waste scarce medical resources. Sadly, both of these scenarios do sometimes occur in countries that allow for active euthanasia. But what is more common is for family and friends to surround their sick and dying loved one, highly distraught at seeing them suffer. Many people, as their life draws near its end, will say that they are ready to die. For some, they are not only ready, they are desperately wanting the end to come. In these situations, a family may decide that the administration of a lethal injection is the most humane and loving thing that can be done. It is in accord with the wishes of their loved one, and they no longer want their loved one to exist in debilitating pain. This does not *justify* the act, but we are dishonest if we say that the motives are the same as those of a cold-blooded murderer. Mercy and tender compassion can be misapplied; good intentions and sincerity do not always lead to right ethical decisions. But intentions do need to be weighed when the moral status of an act is being evaluated.

The question of euthanasia reveals the importance of worldviews and ethical systems. Those who support active euthanasia in the

Western world often point to families who keep dogs and cats as pets. When these pets get to a certain point of sickness and pain, they are humanely put to death. We recognize that it is better to actively bring about the death of our suffering animals than to continue to allow them to live in pain. In other words, when their quality of life deteriorates past a certain point, we think it is more compassionate for us to kill them than to allow them to live. Why would we be more compassionate with animals than people? Surely, proponents of active euthanasia argue, it is immoral to treat animals with more rational kindness than that which we give to our human loved ones. If we give pets "death with dignity," why would we refuse to do the same for people?

What do we say in response to this? The major issue is the *unique nature of human beings*. If atheism is true and if there is no real purpose or point to human life—if we are the accidental outcomes of a mindless process—then what does it matter if we live or die? If we are merely animals, then we can be treated like animals. (However, this is a principle that one might want to think long and hard about before agreeing to implement. Is that really a governing principle that we want to see put into effect in our world?) If there is no God, then we are not ultimately answerable to anyone. We have no obligation to live. Given atheism, then, it would seem that people should be able to take their own lives if they want. If someone believes that their quality of life does not make their life worth living, what right does anyone have to stop them from ending it? And if a medical professional is willing to assist them in fulfilling that desire, then so be it. In atheism, there are no real grounds for the sanctity of life.

> **In atheism, there are no real grounds for the sanctity of life.**

In Christianity, however, life is sacred. There is a God to whom we must answer. Our lives, minds, spirits and bodies are not our own. Human beings are the image-bearers of God, and we are not to bring about our own deaths. Furthermore, we are responsible for upholding God's moral law, which forbids killing the innocent. A Christian, therefore, cannot support murder in any form, even if it is grounded in compassionate motives. In consequence, a Christian cannot support active euthanasia. But, in compassion, love and with deep reverence, a Christian in some circumstances can support unhooking a person from artificial medical machinery and allowing nature to take its course. There is no moral principle that requires us to keep a heart beating no matter what. If the heart would stop naturally apart from the machine, and if the lungs would stop breathing apart from the machine, and if the person will never be responsive again, then the Christian can endorse removing that person from artificial intervention and allowing them to die naturally.

I, Steve, have met with families and doctors in hospital rooms to determine what ought to be done in such circumstances. After attempting to listen and understand, and after allowing every family member to voice their concerns and opinions—and after carefully hearing the doctor's prognosis—there have been times when I have told the family that if they want to unplug the machine, they are not doing anything morally wrong. I have done so with an awareness of the holiness of the life, the love of the family and the fact that God is the judge. But there is a time to be born and a time to die. We do not actively kill, but there are times when we can stop fighting the inevitability of death and let nature take its course.

## MANAGING PAIN IN TERMINAL CASES

Another relevant issue revolves around the management of pain in the terminally ill. Currently, there are medications that are administered to the terminally ill that are very effective in helping them manage pain. The most effective medications, however, can also end up hastening death; not immediately—and they are not the direct cause of death—but over time they further weaken the body. Is it permissible to give a dying person pain medicine that will manage their pain, but will result in them living for a somewhat shorter period of time? We believe that this can be ethically permissible, for the following reason. The principle of double effect states that an action that produces both good and negative effects can be ethically acceptable if the good effects outweigh the bad effects, and the bad effects are unintentional. When it comes to pain management, then, if a certain medication is necessary to control a dying patient's pain, the intention in giving it to them is to help them with their pain. The fact that this medication will slightly shorten their life is not the reason why

> **The principle of double effect**
>
> An action that produces both good and negative effects can be ethically acceptable if the good effects outweigh the bad effects, and the bad effects are unintentional.

the medication is being given—it is an unintended effect. Active euthanasia *intends* to bring about death immediately. Providing effective pain medicine does not have this same intention, nor does it have the same immediate effect. If someone has an irreversible and terminal medical condition, it seems that relieving their pain is a noble and good goal, and if the treatment means that they live nine days without pain instead of ten days with unbearable pain, the treatment can be ethically justified.

## THE BODY AT DEATH

What should Christians do with the body of someone after their death? Christians legitimately disagree on this question in practice, even when they agree about the attitudes that should characterize our handling of the dead. There is never a time when the body is garbage; it is always to be treated with a measure of respect. Human beings are valuable, both body and soul. The fact that the soul has been severed from its unity with the body does not mean that the human body is devoid of all dignity and worth. The point can be illustrated crassly: mutilating and dismembering a dead body for fun would be rightly seen as a sacrilege. We instinctively know that a human corpse is to be respected, and our treatment of it is still a matter for ethics.

### *Organ donation*

In some countries, people have the option of giving permission for their organs to be taken after death for the purpose of organ transplants. In an organ transplant, a diseased, malfunctioning or damaged organ is removed from a living patient, and the healthier organ from the dead person's body is surgically implanted in its place. If a healthy young person is killed in an accident, all of their organs may be used in this way to save the lives of others. As believers, we are to love one another and even be willing to die for each other. We are to be prepared to sacrifice. If we would surrender our living bodies to death to save someone's life, it seems there is nothing wrong with surrendering parts of our bodies after death for the sake of saving lives. Where possible, deciding to be an organ donor is an ethically good decision. It may not be a necessary decision, but it is entirely acceptable. The reality that life for others may come out of our dead bodies is a good thing. The emergence of life from death is both a beautiful and a natural principle. The entire natural world is dependent on cycles of life and death. As God's image-bearers, if we can use our wisdom to bring life from death, we are acting in accordance with the rhythm of the natural world.

### Care for the body: Burial or cremation?

Christians in different parts of the world have a variety of customs when it comes to disposing of the body. In some hot climates without embalming procedures, the dead body needs to be buried quickly. Where we live in Canada, people who die in the winter are often embalmed, kept in a vault and then buried in the spring, once the snow melts and the ground thaws. In Israel during Jesus' time, bodies were often interred in a cave and then their bones were later gathered into ossuaries (i.e., small stone boxes). Burial was the common cultural practice in Judaism, and it carried over into Christianity. Christians believed that the body was to be carefully laid to rest, awaiting the great time of the return of the Lord and the resurrection of the dead.

Burial has been *normative* for most Christians throughout history, but is it *mandatory*? If someone dies at sea, the common practice has been to release the body into the water. This "burial at sea" is a non-controversial practice. Given the circumstances, there is nothing disrespectful about committing the body to the waters. On land, there is nothing disrespectful about placing a body in a burial cave or committing it to the ground.

**Burial has been normative *for* most Christians throughout history, but is it mandatory?**

But is it permissible for a Christian to opt for cremation? In cremation, the body is placed intact into an exceptionally hot fire. Over time, the body is consumed by the extreme heat and is reduced to ashes. These ashes can then be carefully gathered and preserved. As time goes on, more and more people—including Christians—are choosing cremation over traditional burial. Some Christians, however, struggle to believe that cremation is an acceptable alternative.

We do not want to be overly detailed, but when one considers what happens to the body after death, there is no way to dispose of the body that does not involve its complete corruption and decay. Whether buried at land or sea, human bodies will rot. They are consumed by either aquatic creatures or worms. Even if a body is sealed in an expensive coffin, the wood eventually deteriorates. Bacteria are always on and in the body, and after death they go to work breaking down the matter. Bodily decomposition is an unavoidable reality. Most of us never see what happens to a buried body, so we can think that they remain physically much as they were when they were lowered into the ground. This is simply not the case.

Christians who believe that cremation is an acceptable option tend to think that it is not more disrespectful to burn a body than it is to allow the body to decompose. They think that part of the issue is the timeframe (i.e., burning is quick, decaying in a grave is long), and our knowledge of the body's state (i.e., we see the ashes after cremation, but we don't see the rotted corpse after burial). At death, our body's fate is to return to dust. As God said, "for dust you are and to dust you will return" (Genesis 3:19). In English burial services, the minister at the graveside will often say that they are committing the body to the ground, "earth to earth, ashes to ashes, dust to dust." Many Christians today believe that cremation reduces the body to the organic ashes and dust of its earthly elements, and thus is simply a faster way of fulfilling our fate of returning to dust. What the worms and bacteria do over time, the flames do quickly.

**No matter the manner of death, or whatever happens to the body afterward, nobody will be beyond God's power to clothe in resurrection glory.**

Proponents of cremation also argue that the cremated remains can still be treated with dignity and respect. The ashes of the cremated body can be carefully gathered and buried in the ground or placed in a vault. Some people scatter the ashes in a special place, recognizing that the atoms of all of our bodies will be dispersed throughout the world over time anyway. It is no argument to suggest that God cannot resurrect those who are cremated; he is omnipotent, after all! No matter the manner of death, or whatever happens to the body afterward, nobody will be beyond God's power to clothe in resurrection glory.

Another factor leading some Christians toward cremation is the very high cost of burial practices in some parts of the world. Funerals in the Western world typically cost many thousands of dollars. Caskets and burial plots can be extremely expensive. Beyond the monetary cost, however, there are also growing ecological concerns. It is one thing to bury a body, but it is something else to fill the ground with concrete, steel, treated wood, textile, fabric, paint and lacquer. When bodies are embalmed, they are also filled with chemicals. Perhaps we need to think about these practices with a little more care. In major cities, there are millions of people—where precisely are millions of people in millions of caskets supposed to be buried? The truth is, in some major cities they are literally running out of burial plots. (Some countries have mandated cremation, since there is no land for burial available. For any Christian in such circumstances, they can be assured that cremation is not sinful, even if it is not their ideal.)

There are options, of course. There could be a return to much simpler and more natural burial methods. Caskets do not need to be elaborate, ornate and expensive. A very simple and plain (and small) casket made of natural materials may be more fitting as a symbol for death, as well as pragmatically beneficial. When all of the evidence is weighed, we conclude that Christians have the liberty to decide for either traditional burial or cremation. But we would like to see both decisions weighed more carefully and holistically. It may be justifiable to embalm a body so that people have the opportunity to come and have one last look at their loved one, saying goodbye and feeling a sense of finality and closure. We must respect the mourning and grief of those suffering loss. Death is one of the last remaining human experiences that forces us to confront our humanity and mortality. It is an opportunity to reflect on life's priorities and loves. This time should be facilitated with the utmost reverence.

> **Death is one of the last remaining human experiences that forces us to confront our humanity and mortality.**

Whether a body is cremated, embalmed or buried immediately, the end of life is a time for reflection on life and eternity. As a result, our funeral practices should remove distractions and allow for the beauty and decorum of natural simplicity. Love, mourning, weeping and hope are best facilitated in natural environments; the more artificial the surroundings, the less we are able to enter into the reality of life and death.

## CHRISTIAN HOPE AND EXPECTATION
The words of the apostle Paul have been used to comfort believers for over two thousand years:

> Brothers and sisters, we do not want you to be uninformed about those who sleep in death, so that you do not grieve like the rest of mankind, who have no hope. For we believe that Jesus died and rose again, and so we believe that God will bring with Jesus those who have fallen asleep in him. According to the Lord's word, we tell you that we who are still alive, who are left until the coming of the Lord, will certainly not precede those who have fallen asleep. For the Lord himself will come down from heaven, with a loud command, with the voice of the archangel and with the trumpet call of God, and the dead in Christ will rise first. After that, we who are still alive and are left will be caught up together with them in the clouds to meet the Lord in the air. And so we will be with the Lord forever. Therefore encourage one another with these words (1 Thessalonians 4:13–18).

This passage legitimizes Christian grief and Christian hope. When people die, it is right to mourn. Although death is inevitable, it is a terrible enemy. Christians more than anyone else should recognize the pain of death. We were not meant to be separated from loved ones. Spirits and bodies were meant to be united, not separated. When an image-bearer of God dies, it is almost a blasphemy; it is certainly an outrage. This is most clearly demonstrated by Jesus' response at the grave of Lazarus. Although Jesus had intentionally delayed his travel so that Lazarus would die, as he moves to the scene of death and mourning, the text says that he was "deeply moved" (John 11:33). Then Jesus wept (v. 35). After this, the text says once again that Jesus was "deeply moved" (v. 38). In the original Greek of John's Gospel, this language refers to anger and outrage. The impression given is that Jesus stands before the face of death, looks at the pain and loss of his loved ones, and is righteously indignant. He is angry. A world that was very good is now marred by an invading power that ends every life and separates every loved one.

People were created for life, not death. Death, in the most fundamental sense of the word, is *wrong*. It is itself a great evil, and it is the consequence of evil. Christians should imitate their Lord and Saviour by hating the reality of death. How tragic that a life should end and that a loved one should depart! Deep sorrow and mourning can be the right response to death. And those who grieve should not be rushed out of their grief with pious clichés.

Nevertheless, Paul tells us that we do not grieve as the rest of the human race who have no hope. Our grief is tempered with hopeful expectation. It may even be tempered with joy. This is not because death is okay, it is because Jesus has conquered it. Through Jesus, death is a pathway to God's holy presence. At the moment of death, our spirits go to be with the Lord, which is better by far (Philippians 1:23; 2 Corinthians 5:8).

> **Our grief is tempered with hopeful expectation.**

When a Christian dies, their spirit is ushered into the presence of their Lord and Saviour. Jesus told the thief on the cross that they would be together in paradise that very day (Luke 23:43). The realm of purgatory does not exist. Christians who remain alive in this world can grieve because they will miss their loved ones, but they can also rejoice because their loved ones are in the glorious presence of the Lord. Death itself is no match for the gospel of Christ.

Even though the spirits of the righteous dead go into the presence of the Lord, this is not their final and eternal state. They will always be with the Lord, but not as disembodied spirits. As Paul makes clear at great length in 1 Corinthians 15, Christ's resurrection is the guarantee that we will experience a *bodily* resurrection as well. When the Lord returns, our purified souls will be joined to glorified bodies. The eternal state does not consist of us being spirits in heaven; on the contrary, we will be glorified *physical-spiritual* beings in a new heaven and new earth. God is going to renovate the entire universe, and the holy city of his people will stand on the earth. This is our great and glorious expectation. The apostle Peter expresses it this way: "But in keeping with his promise we are looking forward to a new heaven and a new earth, where righteousness dwells" (2 Peter 3:13). God's new creation is the home of righteousness, and only those who are made righteous in and by Christ will live there. Space forbids an exposition of Revelation 21 and 22, but these chapters should be carefully read and meditated upon. Yes, Christians grieve, but we do so in the face of sure and certain eschatological glory. It is a denial of Christ to grieve without hope in the resurrection.

It is very important that we continue to bear in mind that the eternal state is about Christ. *He* is the glory there, rather than our departed loved ones. As much as it will be wonderful beyond description to see our earthly loved ones without sin or pain—that is, when we finally see them *flourishing* as they were always made to be—this is not what glory is about. Glory is not even about us and our experience and our worship. It is about Jesus. So although we long to be reunited with our loved ones—and what a joy that will be!—the true joy is found in the presence of the Lord. There, as C.S. Lewis reminds us, we will live in love.[2]

This truth has received beautiful expression in Anne R. Cousin's nineteenth-century hymn "The Sands of Time Are Sinking." Anne wrote:

> The sands of time are sinking,
> The dawn of Heaven breaks;
> The summer morn I've sighed for,
> The fair, sweet morn awakes;
> Dark, dark hath been the midnight,
> But dayspring is at hand,
> And glory, glory dwelleth
> In Immanuel's land.
>
> O Christ, He is the fountain,
> The deep, sweet well of love!
> The streams on earth I've tasted,
> More deep I'll drink above;
> There to an ocean fullness
> His mercy doth expand,
> And glory, glory dwelleth
> In Immanuel's land.
>
> O! I am my Beloved's
> And my Beloved's mine!

---

[2] C.S. Lewis, *The Great Divorce* (Glasgow: Geoffrey Bless, 1946), 103.

He brings a poor, vile sinner
Into His "house of wine;"
I stand upon His merit,
I know no other stand,
Not e'en where glory dwelleth
In Immanuel's land.

The bride eyes not her garment,
But her dear Bridegroom's face;
I will not gaze at glory
But on my King of grace;
Not at the crown He giveth,
But on His piercèd hand;
The Lamb is all the glory
Of Immanuel's land.

## CARE FOR THE DYING

In history, Christians have excelled at providing care for the dying. During many of the worst plagues in Rome, pagans noticed that as they abandoned the cities and the masses, Christians sought out the sick and dying to provide them with comfort and care. This could be done, in part, because Christians lived their lives anchored in the prospect of eternity. Physical death wasn't (and isn't) the final word on life. Since their bodies and souls would experience resurrection glory in the future, they could risk the possibility of contracting disease and dying as they cared for the suffering. Christians not only died well, they laboured in love to help others die well, too. Offering primitive health-care in the form of a drink of water, a blanket or a morsel of food, believers engaged the dying with compassion, mercy, strength and love. Western hospitals and modern health-care is rooted in Christian soil. Believers are to be servants in both life and death.

Love is indispensable in the provision of proper medical care. Expertise and technology are wonderful in their own sphere, but Christians must draw close to the sick and dying as persons to persons. Love must be concerned with form and structure, since structures are necessary for maximally effective care. Thus, Christians will care about the type of health-care that their culture and society provides. Communities need to be places for all of life, from birth to death. Special care and provision should be made for those who are dying. Likewise, special thought should be given concerning the needs of those who are ministering to their dying loved ones. Death can be met with the dignity of a human touch, and with the purifying power of love.

In many parts of the world, of course, people are born, live their lives and die in their homes. Allowing someone to die at home with the minimum amount of third-party medical intrusion is something that more families in the Western world may wish to consider. We fully recognize, however, that many of the elderly in the Western world never live with their families, and they spend their last years either alone or in nursing homes and hospitals. There are deep structural issues in Western society when it comes to the meaning and role of the extended family, as well as the institutionalization of all health care. In other places, it is the family that naturally attends to the dying. There, birth and death are equally family experiences.

Many cultures have recognized the benefit of having special houses or places for the dying. Although many times people were given the dignity of dying on their own land, special houses or huts for dying people were also frequently used in various cultures and communities. These special places were marked by an orientation to simplicity and the sacred. They represented the opposite of a sterile hospital room. (It must be said, of course, that certain circumstances will result in death taking place in a regular hospital room, and where this is the case it is entirely acceptable.) Where there is the opportunity to construct significant, simple and sacred spaces for dying, this can be a real blessing.

**Where there is the opportunity to construct significant, simple and sacred spaces for dying, this can be a real blessing.**

Death is a major event, and dying in sacred spaces expressly designed for that purpose can be helpful for both the dying and their surrounding loved ones. A hospice ought to provide an environment of simple yet ornate beauty, and a reverent atmosphere. It should be calming. People who enter know that they are in the presence of death and last days. There is stillness. There is lots of room for weeping, but also room for joy and celebration. Ideally, it is a holy place for the final expressions of earthly love. Privacy is protected—this is essential—and yet there is room for all who need to be there. We need to find a way to bring love into an efficient structure so that it can flood out. As the American architect Frank Lloyd Wright very famously suggested, "Form follows function—that has been misunderstood. Form and function should be one, joined in a spiritual union." We need to see the organic, earthy, spiritual and heavenly connections that intersect at the end of life. We need to see God's pattern and design for how we can love those who are dying, and how we can support their loved ones through the process. At least in the Western world, we need to reorient ourselves and our practices to the life cycle of birth, life and death.[3]

## DEATH AND DYING CAN BE SANCTIFYING

We must also remember that God can teach us and develop our characters through every stage and experience of life. Many advocates of active euthanasia fail to see that God can work in the deepest part of a person's life, even in the midst of physical and mental deterioration and pain. In fact, prolonged illness may strip away all of our pretensions and self-trust. It may cause us to cling desperately to Christ, holding on to him and to him alone in ways we wouldn't with full health. The point of life is not to live in happy comfort and pleasure. God has more important things for us in this world than that. We are to learn that we are dependent on God for every breath. Some people who are performance oriented really struggle to accept that God loves them when they are unable to contribute work to the kingdom—but the love of God is not based on our strength. As Paul said, the Lord's power is made perfect in our weakness (2 Corinthians 12:9). In grace, God can teach and refine us even as our lives and powers ebb away. The steadfast love of the Lord is better than life, and his grace is sufficient. It can be easy to say that when we are healthy and strong: Can we rest secure in God's love when we are old or sick?

---

[3] Although we recognize that it is impossible to achieve the ideal in everything, we believe it is best if dying houses or hospices are both easily accessible and semisecluded. Besides the architecture and design of the building and rooms, natural surroundings are important. A grove of trees or a wood with walking paths; a pond or stream; a garden—these are all the kinds of natural beauties that can help put us back in touch with the rhythms of the natural world. When it comes to the great matters of life and death, contact with nature can be healing and anchoring.

The ending of Anne Porter's poem "A Year of Jubilee" answers this question with rare beauty:

> He wishes us
> To be like children
>
> You also told us
> Our Father
> Blesses us most of all
> When we are poor.
>
> So even when our bodies
> Have grown old
> And our heads are filled with confusion
>
> He will not love us
> Any the less for that.[4]

## THE END OF LIFE

We cannot know the meaning of death until we know the end of life. "The end of life" can refer to its termination point (i.e., death), but it can also refer to its *goal*. It is only when we contemplate both meanings of the end of life that we can really live and really be prepared for death. The end of our life in this world is death, but the end goal of life is life! Eternal life. All of our love continues. Forever.

Shepherding people through their final days and hours is a high and holy calling. It is a community calling, and not one that can be merely delegated to professionals in a sanitized and sterile institution. People should die in love. In Christ, the dead depart into something far better, where they await something far, far better. Let those of us who believe love them on their way, and love them when they cross the waters of death. We are left on one shore while they arrive on the other, but one day soon we know that our time will come and we will also cross over. Let us treat others in their death as we wish to be treated when our time comes. Let us remember the end of our lives as well as the chief end of life. Those in Christ will glorify God and enjoy him *forever*.

---

[4] Anne Porter, *Living Things: Collected Poems* (Hanover: Zoland Books, 2006), 98.

## REFLECTION QUESTIONS

1. Do you think there is a morally relevant difference between active and passive euthanasia?

2. What are your culture's traditions when it comes to providing care for the dying? Are they biblical?

3. Do you think that Christians act differently in the face of death than others do? Why or why not? If not, how can we do a better job preparing believers to face death and loss?

4. If you could design a sacred space for the dying in their last days, what elements would you include? What would you want it to be like?

5. Do you favour burial or cremation? Do you think that both are ethically acceptable for Christians? What biblical data supports your view?

# 8

# Ethics in society

Today, much is made of the fact that we are all members of the global community. Technology and modes of transportation have perceptibly shrunk the globe in a way that was impossible to contemplate even just a short time ago. Economics are a driving force behind globalization, as multinational companies look to find resources and make profits in every square inch of the globe. As a result, we tend to be more aware than ever before about the interrelationships that exist between nations. What happens on the far side of the world really does affect us. Many are encouraging people to think in terms of "global citizenship," which includes political, economic and ecological factors. Living in the global village is part of the reality of this century. One does not need to even be aware of these global realities to be affected by them.

We may all be, inescapably, members of the large global community, but we are also members of much smaller local groups. Most people are part of family and kin groups. Widening the circle, there are expanded clan and village groupings. There are local, regional, national and international levels of society, all of which make claims upon us. The authors of this book live in the same city in Canada, and this means that we are also part of the same province, country and continent. There are responsibilities that attend each circle of society. In Canada, there are municipal, provincial and federal taxes, and each sphere of government is supposed to provide certain services.

Being a member of society brings with it many obligations. Sometimes we may love our government, but other times we may think that our rulers are evil and need to be opposed. Regardless of our feelings, however, we live in a political society and world. This reality affects every part of our lives. In fact, it affects us so much that we often simply take it for granted. The political environment is like the physical environment—until there is a crisis, we tend not to notice it. But, every day of our lives, the decisions we make and the opportunities we have are partly sculpted by the society in which we live, our economy, our fellow citizens and a host of other civic factors. Since our lives are lived in society, and since human beings are social creatures, we need to think through some ethical issues that attend this reality. From major political issues—like joining our country's army and fighting in a war—to more normal activities—like going to work for the day—it is imperative that we take our civic duties seriously. To do that, we need to see what God has revealed to us about living in a civil society.

## GOVERNMENT: A SERVANT OF GOD

As we will see, the Bible is very clear that no human government can be our ultimate authority. Rulers can be corrupt or guilty of committing horrendous crimes and atrocities. Apocalyptic and prophetic literature (like parts of Daniel and Revelation) sometimes depict nations and rulers as ravenous and bloodthirsty beasts. The Bible simply does not, however, suggest that all human government and political organization is evil. In fact, quite the opposite. We need to realize that the biblical revelation concerning political government is multifaceted. There can also be a profound difference between what government *ought* to be in theory, and what it actually is in practice, or in particular cases. As is always the case, we must form our views on the

> *As is always the case, we must form our views on the basis of the full teachings of Scripture.*

basis of the full teachings of Scripture. In this fallen world, government has a good function, but it can be corrupted and become a force for evil.

If human governments were *intrinsically* evil, God would not have appointed leaders over his people and organized them into a nation. Yet, right after the Exodus from Egypt, God organized his people into a national entity. He appointed Moses and other leaders to guide and direct them. After Moses' death, Joshua became the leader of the nation. Even before Israel's monarchy, the tribes were organized into a national group. Each tribe had their own leaders, but these leaders were also accountable to Moses or Joshua. In the Book of Judges, one of the great problems for Israel is that there was no central leadership. (Although Israel's monarchy showed that having a king is no guarantee of success either!) God chose to rule his people through the Davidic line of kings. There was nothing wrong with kings *per se*; it was the individual king's faithfulness to the Lord that mattered. Today, there is nothing wrong with government *per se*, but its authority does not supplant God's authority in our lives, and we must be accountable to him above all.

> **[The government's] authority does not supplant God's authority in our lives, and we must be accountable to him above all.**

There are many passages of Scripture that we could look at to support the position that the Bible takes a positive view of human government, but we will restrict our discussion to four crucial ones.

### 1. 1 Peter 2:13–17

> Submit yourselves for the Lord's sake to every human authority: whether to the emperor, as the supreme authority, or to governors, who are sent by him to punish those who do wrong and to commend those who do right. For it is God's will that by doing good you should silence the ignorant talk of foolish people. Live as free people, but do not use your freedom as a cover-up for evil; live as God's slaves. Show proper respect to everyone, love the family of believers, fear God, honor the emperor.

The meaning of this text is clear enough. Believers are to submit to every legitimate governmental authority, and we are to do so for the sake of the Lord. In other words, one of the ways that we obey and honour the Lord is by obeying and honouring lawful human authorities. The emperor is the supreme authority, and governors are those who are appointed to regional posts. This gives divine authorization to honouring government at all levels. Peter gives us a brief description of one of the government's roles: they are to punish wrongdoers and commend those who do right. (We will see that Paul says the same thing in Romans 13.) Notice that in the flow of Peter's argument, one of the reasons we are to submit to human government is so people will not be able to accuse us of being rebellious or antagonistic to the social order. Christians should not be constantly complaining about the government; we are to be the most respectful and well-ordered citizens in the nation. In verse 17, Peter tells us to "honour" everyone and to "honour" the emperor. (The same Greek word starts and ends the list, a fact obscured by the translation "show proper respect.") We are also to love believers and fear God. This last point is very important: Peter began this paragraph by saying that we are to submit to human authority for the Lord's sake, and he ends by reminding his readers that they need to fear God as well as honour the emperor. God is not to be left out of the issues of civil government.

### 2. First Timothy 2:1-4

> I urge, then, first of all, that petitions, prayers, intercession and thanksgiving be made for all people—for kings and all those in authority, that we may live peaceful and quiet lives in all godliness and holiness. This is good, and pleases God our Savior, who wants all people to be saved and to come to a knowledge of the truth.

Paul is obviously instructing believers to pray and intercede for kings and those in authority. Many churches do pray for the government, asking God to save the souls of their rulers and to help them govern with wisdom and justice, in accord with God's Word. It is right to pray that God will give our leaders wisdom—and it certainly is acceptable to pray for salvation—but many times people fail to notice that we are also supposed to pray *giving thanks* for our governments. When Paul wrote this letter, the rulers and authorities were not kindly disposed to Christians, yet Paul still expected his readers to be thankful for them. This is because efficient government is essential to the order and functioning of any kind of complex civilization.

> ...we are also supposed to pray giving thanks for our governments.

Paul links prayer for our leaders with the advance of the gospel. This is crucial. The main reason we pray for our governments is not because we want to pay fewer taxes or have an increase in the services we receive. *Christ's gospel* is the reason we pray, not personal comfort. Paul says that we pray so that we can live quiet and holy lives. In other words, we pray that society will be organized in such a way that we can go about our business and share the gospel freely with the lost. This is a good diagnostic test for politico-theological health: Do we pray for our governingl authorities because we are concerned with our own comfort, or do we pray for our leaders because we are motivated by a concern for the advance of the gospel?

### 3. Romans 13:1–7

> Let everyone be subject to the governing authorities, for there is no authority except that which God has established. The authorities that exist have been established by God. Consequently, whoever rebels against the authority is rebelling against what God has instituted, and those who do so will bring judgment on themselves. For rulers hold no terror for those who do right, but for those who do wrong. Do you want to be free from fear of the one in authority? Then do what is right and you will be commended. For the one in authority is God's servant for your good. But if you do wrong, be afraid, for rulers do not bear the sword for no reason. They are God's servants, agents of wrath to bring punishment on the wrongdoer. Therefore, it is necessary to submit to the authorities, not only because of possible punishment but also as a matter of conscience.
>
> This is also why you pay taxes, for the authorities are God's servants, who give their full time to governing. Give to everyone

what you owe them: If you owe taxes, pay taxes; if revenue, then revenue; if respect, then respect; if honor, then honor.

This is Paul's most important passage when it comes to understanding the role of governing authorities. Thankfully, this passage is not hard to interpret. We are to be subject to the authorities, because they have been established by God. Christians are not to be rebellious, anarchic or even disrespectful. To rebel against the authorities, Paul says bluntly, is to rebel against God. If God has told us to submit to someone, defying them is also defying God's command. Just as Peter did, Paul also mentions that the government is to commend those who do what's right and punish those who do what's wrong. If we do what's wrong, we have every right to fear judicial retribution. Paul explicitly calls the governing authorities "God's deacon" or "God's servant" (the term can be translated either way, but it is literally "deacon.") This is a very positive claim about the nature of government. To function as God's servant or deacon is both noble and good. This is why we are to honour and support the government.

> **Christians are not to be rebellious, anarchic or even disrespectful.**

### 4. Matthew 22:15-22

> Then the Pharisees went out and laid plans to trap him in his words. They sent their disciples to him along with the Herodians. "Teacher," they said, "we know that you are a man of integrity and that you teach the way of God in accordance with the truth. You aren't swayed by others, because you pay no attention to who they are. Tell us then, what is your opinion? Is it right to pay the imperial tax to Caesar or not?"
>
> But Jesus, knowing their evil intent, said, "You hypocrites, why are you trying to trap me? Show me the coin used for paying the tax." They brought him a denarius, and he asked them, "Whose image is this? And whose inscription?"
>
> "Caesar's," they replied.
>
> Then he said to them, "So give back to Caesar what is Caesar's, and to God what is God's."

When they heard this, they were amazed. So they left him and went away.

These words of Jesus have often been taken as of central importance for determining how believers ought to relate to governing authorities. Jesus does seem to endorse a proper sphere for Caesar/government in this text, but that is not the major point. There are multiple nuances in this text, but Jesus leads his challengers to notice that the coin bears Caesar's image. How can you tell what belongs to Caesar? It bears his image. But how can you tell what belongs to God? The answer is the same. Give to Caesar his tax money and the coins that bear his image, but give to God that which bears *his* image. And what exactly bears the image of God? We do. God created human beings in his own image. So we are to give to Caesar what belongs to him, and we are to give *our entire selves* to God. We are to love him with all of our heart, mind, soul and strength (Matthew 22:37). It is wrong to hold back taxes from those to whom they are owed. It is wrong to hold back any part of ourselves from God. This principle establishes both a logical priority of authority, and a hierarchy of ethical importance. Caesar is to receive his due, and God is to receive his due. In the latter case, this is *all* that we are. Thus, if there is a conflict between Caesar and God, our responsibility is clear.

> **Give to Caesar his tax money and the coins that bear his image, but give to God that which bears his image.**

## CIVIL DISOBEDIENCE

The Bible provides positive examples of civil disobedience. There are times in Scripture when a ruler's laws and commands conflicted with God's, and God's people rightly chose to disobey their government rather than God. They were also willing, however, to receive the civic punishment for doing so. One prominent example is recorded in Daniel 3. The king of Babylon built a huge statue and ordered everyone to bow down to it. Shadrach, Meshach and Abednego refused to bow down, and they were thrown into the fiery furnace. God spared their lives for their faithfulness. Under Darius, Daniel disobeyed the king's edict that forbade praying to the Lord, and he was thrown into the lion's den in consequence (Daniel 6). God spared his life for his faithfulness. Another famous example comes from the New Testament. In Acts 4, we are told how the religious leaders in Jerusalem tried to stop the disciples from preaching in the name of Jesus. The leaders commanded the disciples to stop preaching, and, "Then they called them in again and commanded them not to speak or teach at all in the name of Jesus. But Peter and John replied, 'Which is right in God's eyes: to listen to you, or to him? You be the judges! As for us, we cannot help speaking about what we have seen and heard'" (Acts 4:18–20). Putting the data from Daniel together with this passage, it is clear that there are times when we must disobey the governing authorities in order to maintain our loyalty to God. God alone is supreme, and every other authority is derived from him and must be subservient to him.

Among Christians, there is very little debate about this point. It is far more difficult, however, to determine exactly when civil protest or disobedience is an acceptable course of action. There are some countries that ban the preaching of the gospel or forbid worshipping the Lord, and in those cases disobeying the government is clearly the right thing to do. But how often does the government actively try to force people to do something that is obviously wrong, like bowing down to physical idols? What if a government policy permits evil behaviour (eg. abortion) but does not compel evil behaviour (eg. when Pharaoh commanded the Hebrew midwives to put to death infant males)? How can we identify when an actual civil rights issue is worth protesting? And, more controversially, what are morally acceptable *forms* of civil disobedience we should engage in?

**In order to justify civil disobedience, the law must be demonstrably unrighteous and produce evil consequences.**

The way these questions get answered usually depends as much on someone's emotional and ethical intuition as it does on rational analysis. Given God's positive commands to obey the governing authorities, we are not at liberty to disobey a law just because we do not agree with it. In order to justify civil disobedience, the law must be demonstrably unrighteous and produce evil consequences. When a person or group in society is being oppressed by law, and their God-given rights are being denied, then civil protest is justifiable. Legal policies of exploitation or oppression should be combatted. If minority groups are being discriminated against by the governing authorities, the ideal would be that majority groups lead the protest. Given our selfishness, self-interest and urge for self-protection, this rarely happens. Tragically, the government policies that discriminate against minority groups are usually only successful because they are supported by the majority of the population.

Civil disobedience is not the first step to be taken in challenging an unjust law or social policy. Whenever possible, normal legal and political channels should be followed to see if the law or government can be changed. Depending on the nation's political and legal structure (both theoretically and practically), this may include actions such as voting for a different party, lobbying the government, creating petitions and filing legal complaints in court. In some places, none of these strategies are possible, and in other places none of them would be effective.

If these sorts of pathways are either nonexistent, blocked or lead to dead-ends, then stronger forms of protest may be permissible. In America, during the civil rights movement, one common form of protest was the organization of rallies and marches. These mass gatherings attracted public notice and attention to the cause. There were also boycotts. Since people had the right of public assembly, these gatherings did not actually constitute civil disobedience. Civil disobedience came when people broke the law. For example, in America there were laws that enforced racial segregation and discrimination. African-Americans were made to sit in the backs of buses, and they were forbidden to eat at the same counters and drink from the same fountains as Caucasians. Given the blatantly discriminatory nature of these laws, some African-Americans began to refuse to abide by them. They publicly broke the law in order to call attention to the unjust nature of the regulations. In Nazi Germany, there were German families that broke the law by hiding and protecting Jews in their homes. This was also an act of civil disobedience, and it literally saved lives. When the government is enforcing laws

**When the government is enforcing laws that are unethical, evil and harmful, various measures of civil disobedience are acceptable.**

that are unethical, evil and harmful, various measures of civil disobedience are acceptable.

Perhaps the most controversial question concerning civil disobedience is whether or not force or violence are ever acceptable in the fight against an unjust law. In other words, must civil disobedience always be peaceful? Answering this question actually pulls together a number of tightly interrelated moral, philosophical, and theological issues. If someone is a pacifist, then the answer will obviously be that violence is never acceptable in any circumstance. If someone is not a pacifist, however, matters may not be quite so simple, even if they also end up concluding that violence has no place in legitimate acts of civil disobedience. Furthermore, if someone believes that violence may be a legitimate tool in the fight against social injustice and oppression, there are qualitative issues to consider as well (i.e., how *much* force is acceptable?). We will come back to these issues, but it will be helpful to step back first and look at the government and war.

## WAR

It is vitally important to recognize that Christians throughout church history have taken different views about the legitimacy of warfare. Christian pacifists believe that no nation should ever fight in a war, even if they are attacked by an invading army. Other Christians think governments have the right to wage war (in some circumstances), but Christians must never be involved in active combat where they may kill another person. In this model, a Christian may be able to serve in the ambulance corps, but not in the infantry. Christians are seen to belong to the kingdom of heaven, whereas the citizens of the kingdom of the world will fight as the world does. Another common position is that the government has the right to wage war (in some circumstances), and Christians also have the right to participate as active combatants in these just wars. More strongly yet, some think that Christians not only have the right to fight in a just war, they have the moral duty to do so if called upon by their government. Although we may take a definite stand for one of these positions, we must do so with respect for the consciences and godliness of our brothers and sisters who disagree with us.

What every Christian should agree with is the proposition that war is always the result of sin. In a world without sin, the possibility of war is logically excluded. It is logically impossible that every side participating in a war is justified. That is a very sobering truth, especially when we consider how quick we are to justify ourselves. War is always a terrible thing; it is unspeakably terrible. The amount of violence, destruction, death and suffering that is generated by war is beyond human understanding. When we think about all of the good that could have been done if the human and economic resources that have gone into war—and the readiness for war—had been channelled into other areas, we realize that we cannot imagine how much potential good has been lost. Even if fighting a war is ever justified, war is so awful a thing that it must be done with deepest regret, and only as a last resort.

*In a world without sin, the possibility of war is logically excluded.*

### 1. Pacifism

Christian pacifists argue that Jesus' ethic of love is incompatible with violence. They also cite Jesus' words in the Sermon on the Mount, "You have heard that it was said, 'Eye for eye, and tooth for tooth.' But I tell you, do not resist an evil person. If anyone slaps you on the right cheek, turn to them the other cheek also" (Matthew 5:38–39). Further evidence comes from Jesus telling his disciples to put away their swords (Luke 22:38), although this text is quite ambiguous, and it follows Jesus telling his disciples to sell their cloaks to purchase swords (Luke 22:36). The fact that God commanded the Israelites to wage war in the Old Testament is problematic for pacifism, but the pacifist argument is usually that Jesus has replaced lower law with a higher ethic based on love and nonviolence. Most pacifists maintain that war may have been permitted for Israel, but it is not permitted for Christians.

### 2. Some wars are just

Those who believe that it may be permissible to fight in some—not all—wars, believe that the pacifist position is wrong. First, they see a very small amount of evidence for pacifism. Second, they believe that perhaps the key text (Matthew 5:38–39) is speaking of personal insult, not national invasion. As an individual I am not to strike someone who insults me, but if a person is about to kill everyone in my neighbourhood, that is a very different circumstance. Positive evidence for the view that fighting in some wars is permissible comes from the fact that God *commanded* Israel to wage war (which God could not have done if sending his people to war was evil). When certain Roman soldiers asked John the Baptist what they ought to do in keeping with repentance, he told them to be content with their pay and not to rob the innocent—he did not tell them to leave the army (Luke 2:14). Jesus also interacted with a Roman centurion and said that he had greater faith than any in Israel (Matthew 8:5–13). Now, it is an argument from silence to say that Jesus approved of his vocation, but there is no indication in the text that being part of the Roman army hindered this man's faith, or that he needed to change his job. As we've already seen, Paul states in Romans 13 that the governing authorities bear the sword. The sword is a symbol of death. Today, if someone said God has entrusted the governing authorities with guns, nobody would interpret that to mean that the "guns" were not really guns, but rather symbols of nonviolent authority and justice. The Book of Revelation, although highly stylized and symbolic, uses incredibly graphic images of God waging war and shedding the blood of his enemies in battle.

### 3. What criteria are used to decide if a war is just?

It is one thing to believe that war can be justified in theory, but it is something else to know when a situation has arisen that permits a just war. In fact, one could believe that war may be justified, and yet believe that no actual war in history reached the proper standards of justification. Although there is plenty of room for nuance, the basic criteria for just war theory includes the following points. First, war must be a last resort after all other diplomatic channels and sanctions have been tried. Second, the nation's official government must proclaim the state of war. Third, there must be a reasonable chance of success. Fourth, there must be the greatest attempt possible to differentiate between combatants and noncombatants. Fifth, the strategy of war should be to minimize collateral damage. Sixth, the war should be defensive—it is impossible to be justified in going to war if you are the aggressor. (This latter point brings up the issue of whether or not a nation can be justified in making a pre-emptory strike in the face of a serious and imminent threat. Just war theorists disagree if such a first-strike can be categorized as "defensive.") A further distinction can be drawn between a defensive war for your own sake and a defensive war for someone else's sake. For example, if people are being butchered by an invading army on the far side of the world, can a nation justify intruding with their armed forces to stop the slaughter? From a Christian perspective, it would seem that if we can justify defending ourselves, then we can also justify defending others, wherever they are.

**Criteria to evaluate whether a war is just**
1. All other diplomatic channels and sanctions have been tried.
2. The nation's official government must proclaim the state of war.
3. There must be a reasonable chance of success.
4. There must be the greatest attempt to differentiate between combatants and noncombatants.
5. The strategy of war should be to minimize collateral damage.
6. The war should be defensive.

### 4. World War Two: A test case

At this time in history, the Second World War provides the most plausible example of a just war. The Nazis were aggressors, invading nations without just cause and liquidating civilians they deemed undesirable (such as Jews). Political appeals were made, but to no avail. Faced with the reality of the Nazis, even the philosopher Bertrand Russell, who was a vocal pacifist during the First World War, concluded that nothing but a greater force would stop the Nazis from destroying the world and murdering tens of millions of innocents. It seems that there is nothing unethical about stopping a murderer who is about to kill your neighbours. In fact, it is hard to claim you love them if you stand by and do nothing. This does not mean that every action in the war was justified. Atrocities were committed by both sides, and not every

Adolf Hitler's murderous campaign during the Second World War resulted in the deaths of millions of people. Stopping him was one of the key objectives of the Allied Forces.

bomb was dropped on ethically justifiable targets. Still, if one can take an honest look at World War Two and remain a pacifist, then one will likely not believe that any war is justified. If one believes that the war was justified—although terrible—then one is not a pacifist. Studying this concrete example may be very helpful for clarifying one's position (probably even more so than studying the arguments of philosophical ethicists).

## 5. Nuclear weapons

Even among proponents of some kind of just war theory, the issue of nuclear weapons is controversial. Again, distinctions can be made between the theory of having and using nuclear weapons and the way that nations have actually accumulated and used nuclear weapons in history. This is simply not the place to evaluate the nuclear policies of the United States and Russia during the Cold War era. It is also not the place to bring in wider concerns about nuclear power in general, including its production, storage and the disposal of nuclear waste.[1] The immediate question we are concerned with is whether the use of nuclear weapons is ever justified.

It should be noted that many major military advances in history were said to be so terrible that people would not be able to use them. This was the case with certain bow and arrow technologies, the invention of the machine gun and now nuclear weapons. Those who are against nuclear weapons tend to point out that having weapons which can destroy the entire population of the world is not a wise thing. They point out that such destructive weapons make it impossible to differentiate between civilians and combatants, and therefore fail to satisfy just war criteria. The long-term fallout from nuclear weapons is also entirely different from traditional explosives.

Proponents of nuclear weapons tend to point out that the use of two atomic bombs in Japan brought World War II to an end, potentially saving millions of additional lives. They also observe that damage done by traditional bombs was exponentially greater than that done by the atomic bombs. In a world where evil regimes can have nuclear arsenals, it may be prudent for other nations to have nuclear weapons for strategic

---

[1] These issues must be engaged in order to have a fully formed ethical view, but pursuing them would take us too far afield at this point.

deterrence. It is also possible to make small nuclear warheads which can be used with precision on military targets. No matter which position one takes in regard to nuclear weapons, their existence and use is the most difficult issue for just war theorists and people of conscience today.

### 6. Combatants and noncombatants

In modern warfare, separating combatants from noncombatants can be very difficult. If civilians are working in factories that make steel which is then used in the war effort, can they be classified as combatants? What happens when governments bring civilians into military targets as "human shields" so that destroying the target also results in the death of innocent people? Maintaining the distinction between combatant and noncombatant is not as easy as it may seem at first glance. This is especially the case when soldiers dress like civilians, or civilians act as suicide bombers. According to just war theory, noncombatants should be protected as much as possible. Certainly innocent civilian populations must not be targeted. The goal in war must never be decimation; whenever possible, lives must be saved. Exactly what this looks like in real warfare may be virtually impossible to decipher, but it is a principled commitment that must be held.

### 7. Child soldiers

Using children as soldiers is grotesquely evil. Children have been forced into armies in Rwanda, Uganda (with the Lord's Resistance Army) and in other parts of the world. They have often been forced to commit atrocities—sometimes being compelled to kill, dismember or rape their parents or siblings—and are often sexually assaulted and beaten. They are given drugs which lead to addiction, so that they can be controlled by fear, guilt and narcotics. There is no debate about whether the use of child soldiers can be justified; it cannot. We include this section not because it is the subject of ethical debate, but in order to highlight the problem. We must do everything we can to end this horror. International programs need to be established and supported that help rescue and rehabilitate these children. As Christians, we must remember that only the redemptive power of Jesus Christ can heal a child who has been abused this way.

> **Using children as soldiers is grotesquely evil.**

## BACK TO CIVIL DISOBEDIENCE

If someone believes that war can be justified, then they have to believe that there are times when violence is ethically acceptable (even if never ideal or desirable). In just war theory, it is the recognized government that issues the decree of war. But by definition, the government can't issue a decree against itself for civil disobedience! Could a government ever become so cruel to its citizenry that living under their regime was as dangerous as living in a war zone? It would seem so. If that's the case, could people be justified in organizing a rebellion or revolution to replace the government? Some Christian ethicists say "yes" and some say "no." Here is one thought experiment to test the intuition: If you believe that it was right for the Allied armies to fight against the Nazis, would it have been right for German citizens to fight against Hitler, too? If Nazi soldiers came to round up and execute Jews, would it have been right for the German population to defend the Jews with armed force? Why would it be okay for English citizens to use arms against the Nazis, but not for German citizens? God, rather than any government—even our own government—is the lawful authority. Violent revolution in a society would always be terrible, but is it always immoral and unethical? These are questions for Christians to think through carefully.

## CAPITAL PUNISHMENT

In some countries, capital punishment is never used and is virtually never thought about. In other places, capital punishment is a normal judicial penalty, and it is also rarely thought about since its existence is widely accepted. There are other places, however, where debates rage about the ethical status of capital punishment. Christians can be strongly committed to different views regarding judicial execution, just as Christians take different stances toward war. There are actually several principles that connect the topic of capital punishment with the topic of war. A major one is whether people are ever justified in using lethal force to take someone's life. If not, then warfare and capital punishment are immediately ruled out. But if God has permitted the taking of life in certain circumstances, then capital punishment becomes a theoretical possibility.

When we turn to the biblical data, there can be no doubt that God allowed capital punishment. In fact, he commanded it. God prescribed the death penalty for numerous offenses in his old covenant law. The commandment, "You shall not murder," does not contradict these other laws. Murder involves taking the life of the innocent, but God clearly distinguishes between murder and the judicial execution of the guilty. Still, the church is not under the theocratic law covenant of Israel (and neither is any nation today). Of greater significance than the covenant with Israel—where capital punishment is in fact commanded by God—is the covenant with Noah. In Genesis 9:5–6, God says,

> And for your lifeblood I will surely demand an accounting. I will demand an accounting from every animal. And from each human being, too, I will demand an accounting for the life of another human being.
> "Whoever sheds human blood,
>     by humans shall their blood be shed;
> for in the image of God
>     has God made mankind."

In this passage, God himself declares that a murderer should be executed. What is critical about the applicability of this statement is that it is a general covenant with creation, rather than a command to the nation of Israel alone (it predates the Mosaic Law). Those who support capital punishment point out that people are still the image-bearers of God, and God's words in this passage seem to have timeless applicability.

In the New Testament, appeal is also made to Romans 13, where Paul says that we are to submit to the governing authorities, and that they bear the sword (i.e., a symbol of life and death). Jesus submitted to the death penalty that was passed by Pilate (although this is a special case). There is no clear verse in the New Testament that challenges legal capital punishment. Jesus' ethic of love does not necessarily contradict it, especially if loving our neighbours requires executing murderers. Again, the fact that God commanded capital

**The fact that God commanded capital punishment for a variety of crimes in the old covenant law shows that it is not intrinsically evil in every circumstance.**

punishment for a variety of crimes in the old covenant law shows that it is not intrinsically evil in every circumstance.

Concerns are often raised that capital punishment is sometimes used against an innocent person who is mistakenly thought to be guilty. If this happens, there is no possibility of undoing the mistake. This is a powerful emotional argument, and it needs to be taken seriously. Condemning the innocent—even if mistakenly—is a terrible thing, no matter what the punishment. God did, however, command capital punishment knowing that sometimes mistakes would be made. In fact, God knew that Jezebel would arrange for the innocent Naboth to be put to death by framing him and misusing the blasphemy laws that prescribed capital punishment (1 Kings 21). God's law mandated multiple witnesses in capital cases, and this shows that the standard of evidence in a capital case must be very high. Every care should be taken to ensure that justice is done.

A similar objection is that capital punishment can be disproportionately assigned to minorities. This argument is heard frequently in the United States. Statistics can be cited that indicate that ethnic minorities are far more likely to be executed than Caucasians when they are tried for capital crimes. There are various ways of assessing this data, however. One could presumably argue that the solution is to be more even-handed in assigning capital punishment, with the result that more Caucasians receive the death penalty than before. Logically, an increase in the application of the death penalty might be warranted.

**Another way forward would be to keep capital punishment as a legal option, but work hard to fix the biases in the judicial system that generate the disparity.**

Another way forward, however, would be to keep capital punishment as a legal option, but work hard to fix the biases in the judicial system that generate the disparity. If God has ordained that capital punishment is the right consequence for murder, then the problem isn't the penalty, the problem is with the legal system that is using it. Reforming a biased judicial system does not necessarily require abolishing the death penalty. Nevertheless, it is possible to imagine a society in which capital punishment was so unfairly applied that it would be best to stop applying the death penalty, at least for a time.

There is a vibrant debate about whether or not capital punishment is an effective deterrent. Some are adamant that statistics prove that it's not, whereas others are equally adamant that it would be an extremely effective deterrent if it was applied more widely and consistently. From a biblical perspective, the deterrent value of a punishment is not what justifies its application. When God spoke to Noah, he did not say that the death penalty should be used because it's a deterrent, he said that it should be used because the murderer has taken the life of one of God's image-bearers. Perhaps we are to infer from this rationale that a murderer should die because they have "killed" God in physical form. If you "kill" God, then you not only deserve to die, you will die. Regardless of the accuracy of this inference—which is not certain—the Bible does not propose capital punishment *on the grounds* that it is a deterrent. We are quite confident that Christians will continue to disagree on the abiding ethical validity of both war and capital punishment for a long time. As a result, we must listen respectfully to one another and love each other in Christ, even as we seek to understand what the Lord requires and desires.

## ORDINARY LIFE

So far in this chapter we have laid a biblical foundation for the positive nature of government, and we have looked at the hard cases concerning civil disobedience and issues of life and death (war and capital punishment). Although some people do live in the midst of violent rebellions and war zones, most people experience life in the relative peace and calm of an ordered society. What are our ethical responsibilities for normal social activities like working at our jobs? How should we view economics and money? What are our ethical responsibilities to the poor and vulnerable? Many of us might never have to make a decision about whether to support our government in war, or whether we ought to defy our government in an act of civil disobedience, but we all have numerous daily social tasks in which we must engage.

### *Work*

The nature of work is one area of life that has been distinctly affected by sin. There are undoubtedly a variety of experiences of work, from absolute energizing joy to drudgery to oppressive slavery. But it's safe to say that in everyone's life, there will be a time when one works and finds it *hard* and often dissatisfying. Sometimes we blame ourselves for this. Sometimes we blame others. But the reason that it is not easy to work with constant enjoyment and satisfaction is because of the effect of sin on the world. We experience frustration, lack of fulfilment and fruitlessness. We are able to envision—and we deeply desire to accomplish—so much more than we are actually able to achieve, due to lack of ability, scarce resources, human competitors, environmental factors and a variety of other circumstances and barriers.

At the same time, work *is* good. God instructed Adam and Eve to work the land in the Garden of Eden and to make it fruitful, prior to the entrance of sin into the world. Work was supposed to produce fruit in proportion to the effort it required, but because sin has marred all things, work itself has become a burden. "By the sweat of your brow you will eat your food…" (Genesis 3:19) and thorns and thistles come up from the same ground as the good plants of the field. We cultivate, and weeds also grow. But through working well, we exercise discipline on our body, mind and spirit, thereby refining our characters to look more like Christ.

**Work is dignifying because it reflects the image of God in us.**

Our propensity to work is a function of our being made in the image of God. God worked to bring about creation and then designed it with the potential for cultivation so that we can work to direct and structure it. (If we were made in the image of a creative God but were not given any material to work *with* that would certainly make it hard for us to live up to our calling as image-bearers! So, work is *good*.) Work is dignifying because it reflects the image of God in us.

It is clearly important that work is an integral part of who we are as humans, but why else do we work? We do so to contribute to the flourishing of the world, to be productive members of society and to do *good* works. We also work to meet our own needs and wants. This can be as direct as subsistence farming and homemaking or more indirectly through making money to pay for our expenses. Work also provides us quite literally with occupation—it occupies our time. It gives purpose and direction to our days and enables the use of our time to be fruitful. Fruit requires cultivation, which is work. We are also God's instruments for good in the world, and therefore he works through us as we fulfil his calling for us. The word *vocation* literally means *a call*.

## 1. How should we work?

The *manner* in which we work, therefore, must be important, particularly for Christians. So how should we work?

(a) *With skill*: The craftsmen that God wanted to build the tabernacle (and temple) had to be "skilled." Not that we are all required to be exceptionally talented in what we do, but it does make sense that we should do everything to the best of our ability, all things considered. "Competent work is a form of love."[2] All of our abilities and gifts are given by God, and using them well is a way to honour him.

(b) *With balance*: There is a danger of both underworking and overworking. The former tends toward sloth and idleness, and the latter toward stress and chaos. Both of these disrupt the pattern of work and rest as laid out by God and impact many other areas of our lives as well. We need to remember that rest is important to good work, and it is important to who we are as humans.

In the account of creation, God worked for six days and rested on the seventh. His direction to his people was to do likewise: to work and rest proportionately. A day of rest is not a sign of weakness or unproductivity. It is a sign that we recognize our proper place in relation to God—we depend on him for making our work fruitful. It is not by the work of our own hands that we achieve success: "Unless the Lord builds the house, the builders labor in vain" (Psalm 127:1). We also honour him by consciously observing the fullness of Sabbath, which is ultimately rest in Christ. We are not slaves to our labour. The gospel gives us a reason to be passionate about our work, and also a deeper kind of rest. Both reflect our freedom in Christ. We are set free from the bondage of sin to "work as unto the Lord" (Colossians 3:23).

> *There is a danger of both underworking and overworking. The former tends toward sloth and idleness, and the latter toward stress and chaos.*

---

[2] Timothy Keller, *Every Good Endeavor: Connecting Your Work to God's Work* (New York: Penguin, 2016), 70.

(c) *With love*: People are more important than projects, and we should never sacrifice others to our desire to excel in our work. Our motivation must always be to love God and find our identity in him, and to love our neighbour. We should display Christ-like character as we work, and "let your light shine before others, that they may see your good deeds and glorify your Father in heaven" (Matthew 5:16). As we do so, and as we exhibit the fruit of the Spirit, we are reminded to submit our work to the will of God. As we act in virtue, and as the Lord decrees it, we can humbly accept either success or loss. If our work is for him, we will see it as his directing hand on our lives.

*2. Who is our work for?*
This brings up the question of *who* our work is for. Colossians 3:23 says, "Whatever you do, work at it with all your heart, as working for the Lord, not for human masters." Is this not the greatest motivation to do good work that one could look for? God sees all of our acts of service and the heart behind them, and has an eternal inheritance as our reward, whereas our human employers can only offer us temporal rewards at best. If we stop looking for rewards from others to give us a sense of accomplishment, we can look to God and judge our work against his standards. Thankfully, God is also extremely gracious when we fail, encouraging us to try again, and even empowering us by his Spirit to do well as we seek to honour him. Working for God also means that we work for others, out of love. We are God's instruments in this world to benefit others both materially and spiritually. Sometimes this will require great sacrifices, but a life exhausted for others is in line with Christ's example on this earth.

*3. Are there types of work we should not do?*
Are there types of work that a Christian ought not to do? We think the answer is an obvious "yes," but with some nuance. As we mentioned earlier, we are primarily under the authority of God. Anything we are required to do or are asked to do by an employer must not contravene God's laws and ethics. It is certainly unethical to engage in any work

that directly promotes or creates immorality. "I'm just doing my job" is never an excuse for acting immorally. This attitude has perpetuated harmful situations around the world for too long. I, Danielle, work as an architect, and it was impressed upon me early on in my education that it was an architect that enabled the mass murder of Jews during the Holocaust by designing the gas chambers. The architects knew exactly what the purpose of the buildings was, and methodically created them to serve that purpose as efficiently as possible. Doing their job indirectly resulted in countless murders of innocent people.

Work that promotes exploitation and oppression is likewise unethical. There is no justification for knowingly performing work that is intended to harm people. However, unknowingly harming people can also be unethical. Negligence is a failure to use proper care which results in damage, and is criminally punishable in many societies. A worker who seeks to exercise skill in all they do should work hard to avoid making these mistakes. However, there are cases in which inexperience or lack of necessary resources results in negligence, and while the person may be held technically responsible, their failure may not necessarily be considered an unethical one. Sometimes we do our best, but what we produce is insufficient, or we make mistakes. Although some mistakes do rise to the level of moral failures, this is not the case for every error.

There are other types of work in which the end result is ambiguous or unknown. We don't believe that it's immoral for a factory worker to assemble cars that may one day kill someone in an accident. There are just far too many factors involved, and it is not the car itself that causes the death, nor are cars produced for that purpose. If the accident was the result of a faulty part, the company would certainly have to take responsibility for the damages, but mistakes will be made and the intent of the worker was not to cause harm, even if the harm resulted from poor workmanship.

### *Finances and stewardship*

Out of all of the sections of this book, this one on financial ethics proved the hardest to fit in. We wrestled with whether this topic should be considered in a full chapter on its own, or as a smaller section linked with life in society. Clearly, we chose the latter option. This is not because there isn't a lot to say on this subject, or because it's unimportant.

Neither is it because the Bible doesn't have much to say about money and economics. In the end, we decided that there are basic biblical principles that are easy to understand, and so our treatment of this topic would be simple. There are certain attitudes and dispositions that God requires of us when it comes to our money, and these can be understood by everyone. What follows, then, is not a prescription for how to organize an economy: it is a short discussion about the attitudes that followers of Christ ought to have when it comes to finances and material possessions.[3]

One of the reasons why the Bible's teachings on possessions are simple is that they must be universally applicable. Christ's teachings are authoritative whether we live in an agrarian society where most people live directly off the land, or if we live in a highly technological society where most people live in cities. A corporate executive with a huge salary needs to go to the Bible to find out how God expects them to steward their resources, but this is no less true for the subsistence hunter-gatherer. Of course, different people have different vocational callings and responsibilities when it comes to the wider economy. Some people make decisions that have major effects on the entire global economy, whereas others have virtually no influence at that level. Nevertheless, given the realities of global manufacture and trade, many of the readers of this book will be able to make purchasing decisions which can help support either ethical or unethical international business practices. As stewards, we should exercise care.

> **Christ's teachings are authoritative whether we live in an agrarian society or a highly technological society.**

---

[3] We do believe, however, that there is a great and pressing need for Christian economists and financial leaders to help transform local, national and global economic systems. The root of selfishness can be destroyed by the gospel, and other-centred love and care can be put in its place. This individual transformation is crucial, but so is the transformation of economic *structures and systems* so that economic injustices and the wasteful exploitation of natural resources are stopped. Good hearts and balanced systems need to work together. C.S. Lewis rightly said, "If you are not a professional Economist and have no experience of Industry, simply being a Christian won't give you the answer to industrial problems" [*God in the Dock: Essays on Theology and Ethics* (Grand Rapids: Eerdmanns), 36]. We need Christian virtues and priorities, as well as specialized and technical wisdom to address societal and global economics.

When it comes to beginning to assess our responsibility before God in regards to stewardship, perhaps the first thing that we need to get clear is that *everything* ultimately belongs to God. Whatever he entrusts us with is ours only in a provisional sense. We are to exercise care as stewards over the money and goods we have been given. Some Christians seem to think that God requires them to tithe—give 10 per cent of their income to the Lord's work—but then the remaining 90 per cent belongs to them to be used for themselves. Some seem to think that God doesn't care how they spend the remainder! It is far more biblical to consider 100 per cent of our resources as belonging to God. All of our economic decisions should be made with an eye to what pleases God. Economics and financial matters need to be subsumed under the greatest commandment, which is to love God entirely.

**The first thing that we need to get clear is that everything ultimately belongs to God.**

All of our economic decisions must also be made with an eye to the second greatest commandment, which is to love our neighbour as ourselves. How can the rich honour Christ when they fill their houses with luxury goods while their neighbours go without basic necessities? How can those in the Western world love their neighbours when they purchase products produced by deeply impoverished people in foreign countries, just so they can buy cheap consumer goods? Followers of Jesus Christ are to be quick to share their resources with others. God's priority for compassion and care for the poor and vulnerable is made absolutely clear in the Old Testament Law, the Prophets, the Gospels and the Epistles (see particularly the Book of James)![4] But we are also to exercise care for our neighbours in what we consume in the first place. If we can purchase goods which are unjustly produced, this does not make it right for us to do so. If cheap consumer goods are only cheap because the workers are being unfairly compensated, this is something that Christians must stand against. Somebody always pays for overly cheap goods—if it is not the consumer, then the payment is being taken out of the pocket of the workers.

---

[4] We will cite various texts throughout the following discussions. There are far too many to list!

The truth is, it is impossible to participate in the global economy without indirectly supporting unethical business practices. Things are simply far too intricately connected and complex for us to be able to know the history of everything that we purchase. Even then, there are some necessities that we must buy which are the products of unfair economic systems. We can try to minimize our complicity in supporting unfair practices, but there is only so much we can do. Nevertheless, we should do all that we can to support ethical businesses. If there are things that we know are unfairly manufactured, we should go without them if at all possible. Some companies only understand profit margins, and it is only when consumers direct their money elsewhere that the companies will change. Depending on where we live, how much we consume and what kind of access we have to information, we have a responsibility to try to make economic decisions that are as informed as possible. This is simply one more way of loving our neighbours, helping others flourish and keeping each other safe.

> **Some companies only understand profit margins, and it is only when consumers direct their money elsewhere that the companies will change.**

In the Western world (as in many other places), one of the greatest ethical issues in economics is *overconsumption*. Simply put, the world's governing economic philosophy pushes people to buy more and more, but it never asks the questions, *When do we have enough?* When will we be *content*? It is not an overstatement to say that many people are addicted to buying new things. The richest nations in the history of the world are all in debt. In the United States and Canada, many people are in personal financial crises because they spend more money than they earn. What this means is that the people with the highest standard of living in history, the people with the most comfort in history, the people with the most food in history—the people with the most *stuff* in history—are *not satisfied with the amount of things they possess*. Much of our global economy is built on greed, exploitation, fear and pride. This is a blunt assessment, but we believe it is both true and defensible. Christians need to be very countercultural when it comes to our financial priorities, and the role that possessions play in our lives. We must take seriously the words of Jesus, "A person's life does not consist in the abundance of their possessions" (Luke 12:15).

Learning to be content is part of growing in holiness. In Philippians 4:11–13, Paul writes:

> I am not saying this because I am in need, for I have learned to be content whatever the circumstances. I know what it is to be in need, and I know what it is to have plenty. I have learned the secret of being content in any and every situation, whether well fed or hungry, whether living in plenty or in want. I can do all this through him who gives me strength.

It is only by the strength of Christ that we can be content in all circumstances. In fact, it may be more difficult to be content when we are surrounded by plenty than when we are in want. When we have much, we tend to take it for granted and fail to be grateful. This is surprising to many people, but the Western experience proves it to be true. Those with the largest number of possessions are not always the most happy or content.

A biblical economic philosophy does not endorse selfish capitalism or selfish socialism, because we cannot honour God if we are being selfish in any economic system. Greed is not to be a motivator for Christian purchasing and the accumulation of possessions. In fact, Paul warns Timothy that even *the desire to be rich is wicked*. We need to be clear: becoming rich is not inherently sinful, but the *desire* for riches and luxurious affluence is. Paul is very clear:

> **It is only by the strength of Christ that we can be content in all circumstances.**

> But godliness with contentment is great gain. For we brought nothing into the world, and we can take nothing out of it. But if we have food and clothing, we will be content with that. Those who want to get rich fall into temptation and a trap and into many foolish and harmful desires that plunge people into ruin and destruction. For the love of money is a root of all kinds of evil. Some people, eager for money, have wandered from the faith and pierced themselves with many griefs (1 Timothy 6:6–10).

If we take Paul's words seriously, there can be no other conclusion than that the motivating force behind the world's economic theories is antithetical to godliness.

Christians ought to be content. Christians ought to be able to be satisfied with *enough*. Overconsumption is damaging spiritually, but it is also damaging physically, psychologically and emotionally. (So is coveting, even if the covetous person cannot indulge their desires.) Overconsumption requires the exploitation of natural resources in an unsustainable way. When we overconsume in the present we deprive the poor of resources they need, and we also deprive future generations of the resources they will require. The environmental toll of overconsumption is evident, if we take the time to look. Christians need to recover their place as creatures who are tied to God's created order—we exercise dominion, but as stewards, and there are certain created limitations to which we need to gladly submit. Economic practices that harm ourselves, our neighbours and the earth simply cannot be pleasing to God. The reason we are to be content is expressed in Hebrews 13:5: "Keep your lives free from the love of money and be content with what you have, because God has said, 'Never will I leave you; never will I forsake you.'" Christians have *God*, and God will never leave us—he is our eternal home and possession. If we cannot be content with God and eternal glory, what will make us content? Will an increase in fading, temporal possessions really make us happy, if having God is not enough for us? The secret to contentment is loving God and knowing he is ours forever.

**When we overconsume in the present we deprive the poor of resources they need, and we also deprive future generations of the resources they will require.**

Since our world is not a place where we can find healthy attitudes toward possessions, where do we look for our example? We look, of course, to Christ. We not only study his teaching but also look at what he has done for us. The best biblical passage to read on this topic is 2 Corinthians 8:1–9:15. Although the entire passage must be read and carefully applied, we will highlight just one verse: "For you know the grace of our Lord Jesus Christ, that though he was rich, yet for your sake he became poor, so that you through his poverty might become rich" (8:9). Christ is our example. He had everything in glory, but he was willing to make himself nothing and be poured out to death for us. That is our economic example. Not personal acquisition and accumulation, but self-giving love and service for others. In other words, the example that Christ provides for our personal stewardship of our finances is one that is completely opposed to the way that the world views money, goods and consumption. This needs to be taken seriously.

## *Caring for the poor and vulnerable*

We all need to take seriously the biblical priority of providing for the practical needs of the poor and vulnerable. God commanded the Israelites *not* to maximize their own profits, but to leave some of their harvest for the poor: "When you reap the harvest of your land, do not reap to the very edges of your field or gather the gleanings of your harvest. Do not go over your vineyard a second time or pick up the grapes that have fallen. Leave them for the poor and the foreigner. I am the LORD your God" (Leviticus 19:9–10; see also Deuteronomy 15:7–11; Psalm 41:1; Proverbs 22:22–23). Jesus also highlighted the importance of caring for the poor (cf. Matthew 19:21; Luke 14:12–14). Paul reports that although the other apostles didn't try to add anything to the gospel message that he preached, they did have one request: "All they asked was that we should continue to remember the poor, the very thing I had been eager to do all along" (Galatians 2:10). James sternly warns the church about showing favouritism to the rich over the poor (James 2:1–4). To illustrate dead faith, James pictures a person who can help the poor materially but refuses to do so (James 2:14–17). Before this, he declared, "Religion that God our Father accepts as pure and faultless is this: to look after orphans and widows in their distress and to keep oneself from being polluted by the world" (James 1:27). Those who are affluent need to be generous with the poor. God gives people wealth not for their own luxurious self-indulgence, but so that they can experience

**To illustrate dead faith, James pictures a person who can help the poor materially but refuses to do so.**

the blessing of being his channels for sharing love and goods with those in need. Christians have a responsibility to work hard in order to take care of themselves and provide for their families (whenever possible). We are also to care for our neighbours, the poor and the marginalized as opportunity arises.

Christians must also prioritize their finances so that they can help support work that is done in the name of Christ. Churches and missions that proclaim the gospel and do good works require financial support, and Christians who have the ability to assist ought to do so. What better way could we spend our time and money than in helping share the love of Christ in the world, and helping spread the gospel? The world, certainly, is not going to allocate its resources to help win people to Christ. If we have benefited from the generosity of Christ, then we ought to be generous, imitating him. Are we? Is the church marked by a radically different attitude toward possessions than the rest of the world? Let us strive to bring our attitudes and practices in line with Jesus' famous words:

**If we have benefited from the generosity of Christ, then we ought to be generous, imitating him.**

> Do not store up for yourselves treasures on earth, where moths and vermin destroy, and where thieves break in and steal. But store up for yourselves treasures in heaven, where moths and vermin do not destroy, and where thieves do not break in and steal. For where your treasure is, there your heart will be also (Matthew 6:19–21).

## REFLECTION QUESTIONS

1. Do you believe that war is ever justified? Why or why not? If it is justified, what criteria need to be used to make that determination?

2. Is civil disobedience ever justified? Why or why not? If it is justified, what criteria need to be used to make that determination?

3. In your experience, are Christian employees better workers than non-Christians?

4. One of the major emphases in Scripture is on providing for the poor and vulnerable. What passages can you cite that talk about our responsibility to care for the poor? Are you and your church doing enough to care for the poor, either locally or abroad?

# 9

# Environmental ethics

The earth is the common home that we have to share with the rest of creation. It is literally our common ground. The earth belongs to God: "The earth is the Lord's, and everything in it…" (Psalm 24:1; see also Leviticus 25:23 and Deuteronomy 8:10; 10:14). It is vital for us to understand our place in creation and in God's order. We are his tenant caregivers. God charged humans with the responsibility of caring for creation. We are not to exercise unbridled power over it as if we created it, nor are we to put ourselves below it and make it an object of worship. We ought to love it neither too much nor too little. We have no absolute claim to it because it is a gift from God. What conceit to think we own it! We also cannot be fatalistic about the state of nature such that we relinquish all responsibility. We are ethically obliged to act responsibly. We have a moral mandate from God, and

> **We are his tenant caregivers. God charged humans with the responsibility of caring for creation.**

we will be called to account for our actions and inactions.

We must also recognize that we live in a fallen state where all of creation experiences the effects of sin: "We know that the whole creation has been groaning as in the pains of childbirth right up to the present time" (Romans 8:22). Because of this, throughout history humans have been contending with the environment. We have often seen nature as being against us. We have used technology to overcome the challenges it poses, such as weather, climate, animals, insects, disease, etc. We build shelters to protect ourselves from its forces. We construct ships and airplanes to carry us across waters to far-off places. We manufacture vehicles and roads to go faster and farther. We make greenhouses in order to control food production environments. In order to do all of these things, we extract vast amounts of natural resources. But is this what "dominion" is supposed to look like? Have we gone too far in some ways in acting against nature? Does life on earth have to be shaped by contention, or can it involve working with the natural order to produce fruitfulness on the earth as well as within ourselves?

**Have we gone too far in some ways in acting against nature?**

The environment is simultaneously integrated with human life and separated from it. It operates without our prompting to affect us (eg. plants growing from seed without cultivation, or volcanoes erupting) yet it is also deeply affected by our activities (eg. animal husbandry, deforestation or water pollution). God has given order and modes of operation to the world, and he has also given humans agency to affect those modes, for good or ill. The environment itself is not a moral being, capable of making judgements, but we have moral obligations toward it because it belongs to God.

This complexity in our relationship to the natural world has made it difficult to make clear moral decisions in every aspect of that relationship. The deep interconnections between our actions and their global consequences can be almost impossible to trace. We are left with a seemingly endless barrage of difficult questions: What is our moral responsibility in such a complex system? Does every decision we make have to be moral in intention and in consequence? What about things we are ignorant of because they have become so common in certain cultures? For example, in North America many people buy plastic food containers that are only designed to be used once. Some are recycled, but others end up filling landfills and oceans. Do we simply weigh benefits and drawbacks of every option we are presented with, and, if so, can we ever do that thoroughly enough? Are there any clear and simple moral guidelines for how we are to live in the environment?

In Western nations, the amount of "ethical" options available can be overwhelming. In other parts of the world, there are very few options available to choose from. Are ethical responsibilities weighted differently depending on one's ability to access alternatives? Is it morally acceptable for those in North America to drive their cars to work every day, knowing that using and extracting fossil fuels pollutes the environment? But what if housing is too expensive near the place where you work, or there are simply no residential options close enough to your place of employment? Is it morally acceptable for us to purchase bananas that come from a country with low standards of workplace health and safety, where the plantation workers are exposed to harmful chemicals daily so we can eat cheap bananas? Where do we draw the lines? What do our intentions count for? How can we even begin to find out how our countless daily choices affect the environment? In a society that is built on a resource economy, how can we continue to function when every decision we make has environmental implications? What do we do if we can't afford the more ethical options? What do we do if we live in a place where all the resources necessary for life are polluted and seemingly irreparable? And if local, small-scale living presents the most ethical option, is that even viable in some parts of the world today?

We don't need sophisticated scientific analysis to know that human beings have a direct impact on the natural world. Since the Industrial Revolution, this impact has been growing exponentially. We have gotten into the habit of creating complex technological solutions to problems where a simple natural solution may have sufficed. These new solutions often create new problems of their own, which then require further technological intervention. If we would take the time to observe the balancing activities that occur in nature, we might see that sometimes it is best not to intervene in the first place. More often than not, the best solution is for us to alter our own behaviour, rather than altering the

> **More often than not, the best solution is for us to alter our own behaviour.**

environment to adapt to our desires. An environmental ethic must involve an acknowledgment and acceptance of certain limitations on how we interact with the natural world.

We'll start by describing what exactly is meant by "the environment," and what its nature is. Then we'll turn to biblical explanations for what our relationship to the environment ought to be, which will help provide a framework for ethical decision making. This will be explored through the broad categories of land, animals and people.

## WHAT IS THE ENVIRONMENT?

An environment is the surroundings or conditions in which one lives or operates. Our natural environment includes the earth and all that is in it, which was created by God. This means nature (plants, animals, water, air, land, etc.) as well as human beings. While the environment is distinct from us in nature, we also participate in its existence, for good or bad. Today, we can also include the broader context of the universe. We have expanded our environment by our explorations of outer space. Despite the fact that just a miniscule portion of the population actually enters outer space, it has become a part of our area of influence and knowledge. We are also leaving our mark there with "space junk," consisting of satellite debris and spent rocket stages now orbiting earth.

The earth belongs to God and was created by him. It is not divine itself. God made it and declared that it was "good" (i.e., well made). From its inception, it was intended to be fruitful (eg. every seed-bearing plant was to reproduce), but it was in need of development—it required rain to cause seeds to sprout and humans to till the earth. The environment was created with a finely tuned relational balance and to be a place for relationships to exist in. All of the created order is interdependent and interconnected. We have local environments, but these also overlap. What happens in one place affects others. Because the environment is made up of living organisms, it is hard to contain the effects of environmental damage (eg. oil spills, pesticide use, insect infestations, etc.). Environments cross political and social borders, and the management of them requires international cooperation. Because of their innumerable

> **All of the created order is interdependent and interconnected.**

points of connection, we really cannot isolate individual elements, and when we attempt to do so we find that there are repercussions in areas we may not have expected.

## WHAT IS THE VALUE OF THE ENVIRONMENT?

Creation has both *intrinsic* value and *instrumental* value. That is, God created it and saw it as "good" prior to creating humans. Then, he gave every seed-bearing plant to humans to be used for their food (Genesis 1:29). Adam and Eve were instructed to take care of the earth in order to increase its fruitfulness (Genesis 2:15: "to work it and take care of it"). Thus, God created mutually beneficial relationships between humans and nature. Both the environment and humans are valuable in themselves, but they also have instrumental value for one another.

Nature is designed to be fruitful. A reproductive imperative is built into every living thing. In Genesis 1:11, the Bible distinctly says that God created seed-bearing plants, and trees that produce fruit with seed in it. In Genesis 1, whenever God creates a new type of living thing, he gives it a command to reproduce: plants and trees are to bear seed and be fruitful; sea creatures, birds, and land animals are all to be fruitful and increase in number. In preparation for the flood in Genesis 6:19, at least one of each male and female animal was brought into the ark, to ensure reproduction and the continuation of each species on earth.

Every creature has a function within the environment. Nothing is superfluous in creation. From the smallest bacteria to the largest mammal, each serves a purpose in the operations of nature. Scientists have categorized the elements of nature by their functions (eg. the food web, the carbon cycle, etc.), which is helpful to explain the interconnectedness of everything. However, isolating these functions almost doesn't go far enough because everything in creation is multifaceted and part of multiple operational systems. For example, a tree may provide food for certain animals, but it also provides shelter, erosion control, soil building, oxygen, etc. If we rationalize cutting down a tree based on only *one* of these criteria, we unintentionally alter many other ecosystems. If these other systems remain unexamined in the destruction of a forest, we may have some big surprises ahead. Creation exists in *interdependence*—one element built upon another. We can see this

> **Nothing is superfluous in creation.**

from the creation account: God begins with light, then separation of elements (sky, water), land, vegetation (which required land), sun and moon, sea creatures and birds (who required water and air), land creatures and, finally, humans.

There does appear to be a hierarchy of instrumental value, however. The latter creations depend upon the former, in ways that the former don't depend upon the latter. Nothing can survive without sun, water or air. And humans, while we sit at the pinnacle of creation—being made uniquely in the image of God—are also at the "top" of the food chain, and are therefore the most dependent on the well-being of the other elements of nature.

Nature is designed to give glory and honour to God. It reflects his creativity, flourishing, beauty, fecundity and integration of parts and whole. Even though we've discussed the fact that nature has instrumental value, it is important to remember that its actual purpose is not found in what it can *do* for us, but in the fact that *it brings glory to God* (cf. Psalm 148:3–5).

**Nature's actual purpose is not found in what it can do for us, but in the fact it brings glory to God.**

Every creature also reveals elements of God's glory and character, and nature is a continuing revelation of the divine (Psalm 19:1–4). The Bible uses many metaphors from nature to describe aspects of God. Yet no finite, created being can ever reveal or image everything about our infinite Creator. When each thing is seen as part of a whole, in relation to all the other parts, it helps us to understand elements of who God is.

However, we can also see that the environment itself experiences the negative effects of sin in the world and is hindered from thriving by many factors. The land produces thorns and thistles, as well as edible food. Famine, floods, landslides and drought frequently occur throughout the world. The land is absolutely necessary for our survival, but it so often needs our intervention to mitigate destruction from natural and manmade disasters, and to help maintain its fruitfulness. Dysfunctional nature doesn't give an accurate picture of the Creator. But when it functions as it was designed, the environment glorifies God.

## WHAT OUGHT TO BE OUR RELATIONSHIP TO THE ENVIRONMENT?

Humans have been tasked with being stewards of the earth, to work it and keep it. This is a pre-Fall mandate from God and entails a long-term productive relationship. As stewards, we have unique roles and responsibilities toward creation. These can include conservation, preservation, extraction and cultivation. We are to be disciplined in how we act toward the land—to use it, as well as to allow it to rest. What are these roles and responsibilities for? Why do humans have them rather than other creatures?

Humans were created in the image of God. We are moral beings, with minds that think and reason, and we are able to act outside of instinct to both create and destroy. Our lives consist in using the resources surrounding us. It must be understood that because of the value of creation described above, our interactions with it do have moral implications. We are able to reasonably see the consequences of our actions, as well as the virtues required to live responsibly in this world. We have a positive obligation to make the world fruitful; we can help to bring out the life God has embedded in creation. In preserving the integrity of creation and making it to be fruitful, we must remember we are cultivators, and not creators.

> ...we must remember we are cultivators, and not creators.

## DISRUPTION TO GOOD CARE

The earth is under the burden of sin, and is presided over by a fallen people. It is now undeniable that humans have greatly impacted the natural environment in negative ways, such as greenhouse gas emissions, soil depletion, increasing soil salinity, oil spills, species extinction, overfishing, etc. Our sin impacts our relationship to the environment. Cain's sin led to his estrangement from the earth (Genesis 4:9–11). The Israelites' sin led to the Promised Land being withheld from them for forty years as they wandered in the fruitless desert.

The Sabbath day was implemented by God as a part of the process of creation. It was the final and culminating step in the order of creation. Like the Year of Jubilee, a time of rest for the land likely contributed to ecological balance as well as fairness among people who used the land (Leviticus 19:9–10). It also served as a reminder that the land belongs to the Lord. The earth and its fruits are for the good of *all* people, not to be held by some indefinitely, to the exclusion of others.

It is hard to sustain the argument that, in order to attain their own ends, humans are permitted to exploit the earth without regard for the rest of creation. To encourage absolute domination and destruction does not fit within a consistent Christian worldview. Despite the term "subdue" which has been used to justify destructive patterns, a more accurate interpretation of the biblical text indicates a role of stewardship—to "till and keep" the garden God created (Genesis 2:15). This involves cultivation and work, as well as care, protection, preservation and oversight. The earth provides for our needs, but we are only to take what is necessary and not exploit the finite resources. Out of care for others, we ought to leave the world in a condition that allows for future fruitfulness. Even during harvest, the principle of gleaning allowed for the poor to live off the excess of the land. We are stewards for God, our neighbours, ourselves and future generations.

> **The earth provides for our needs, but we are only to take what is necessary and not exploit the finite resources.**

Wendell Berry draws this out of the scene in Homer's *Odyssey* where Odysseus has finally returned home and meets his father Laertes for the first time in many years. Part of Homer's poetic description reads, "Odysseus found his father alone on the vineyard terrace hoeing round a tree."[1] Berry writes, "In a time of disorder he [Laertes] has returned to the care of the earth, the foundation of life and hope. And Odysseus finds him in an act emblematic of the best and most responsible kind of agriculture: an old man caring for a young tree."[2]

We have a positive duty to protect the earth and ensure it continues to provide for future generations. As image-bearers of God, we are to protect life. We must respect the goodness of every creature, and avoid any disordered use of them.

We are not God. We cannot wipe the slate clean and start again whenever we want. We could never replace God's complex and miraculous creation with our own creations. If we destroy the foundations

---

[1] Homer, *The Odyssey*, trans. E.V. Rieu (Toronto: Penguin, 2003), 316.
[2] Wendell Berry, *The Unsettling of America: Culture and Agriculture* (Berkeley: Counterpoint Press, 1977), 133.

and conditions of life for a species or an ecosystem, we are simply not capable of recreating it and we lose something irreplaceable. If God considered something worth creating, we are not to disregard it as expendable or useless. Countless species have become extinct over human history, and we have not been able to bring them back. Neither have we been able to recover some damaged ecosystems, or clean up all the polluted air and bodies of water. It is not merely a matter of time before our technology makes this possible. It cannot be simply a matter of time and effort before we become "like God," able to create out of nothing. This hubris has always been the fundamental sin, and if we predicate our approach to the natural environment on this error, we make ourselves out to be fools, trusting in the strength of our own hands.

As in other areas of ethical decision-making, it is a proper worldview that is needed—one that supports care for the earth and a holistic, interconnected way of living. This enables the environment and our life within it to function best. Our moral behaviour affects the fruitfulness of the world.

So what *are* the ethical environmental decisions we need to be aware of? We will use three broad categories to help organize a selection of the topics: land, animals and people. Many of these topics are interrelated and can be dealt with under more than one category. There are more topics about the environment than we can cover in this chapter, so we are only presenting a limited selection.

## ENVIRONMENTAL ETHICS AND LAND

Simply put, all land belongs to God and is a gift to human beings. He is the creator and the sustainer of all things. God is the landlord, and we are the tenants. Having been tasked with stewardship, what are our responsibilities toward the land itself?

Water is arguably the most essential element necessary for life. Water is needed for direct consumption by humans and animals, for hygiene, and also indirectly for the crops we produce to eat. One of the most frightening problems worldwide is the prospect of running out of a reliable supply of fresh water. Sixty per cent of the world's fresh water is found in only nine countries.[3] The preciousness (and volatility) of this resource can be illustrated by Cape Town, South Africa,

---

[3] http://www.fao.org/docrep/005/y4473e/y4473e08.htm

where they have struggled with a long-term water shortage crisis, nearly coming to the point of having to literally turn off the taps in 2018.[4]

If we are to be concerned about managing our water resources well, we have to carefully consider both our urban and rural consumption. In cities, it is easy to over-consume water because it is so readily available and easily accessible through infrastructure. Twenty-five litres of water per person per day is required for basic needs. The average North American consumes approximately 197 litres per day.[5] Can this be justified in a world where people are dying because of lack of access to clean drinking water? Even though decreasing consumption in water-rich nations won't directly impact those lacking access to water, conserving the resource now will help to protect against future scarcity. There are philosophical and sociological debates about whether water is to be considered an "ecological common resource" rather than a commodity, and whether access to clean drinking water should be considered a basic human right. Regardless of this discussion, and although there is good reason to see it as a common resource, it is all ultimately God's, and it only makes sense he would want it to be available to all people. More obviously than other ecological resources, water cannot be indefinitely owned or contained—it is part of a cycle that is constantly being refreshed and returned.

**What is a moral approach to the distribution of resources like water?**

What is a moral approach to the distribution of resources like water? Are we to only take our "fair share," and what does that look like? Thoughtful and conservative use would be an obvious response. Between 1999 and 2016, average household water use in North America decreased by 22%. This is largely due to more efficient appliances like toilets and clothes washing machines. Many people are going a step further and using "grey water" systems, which recycle household water to be used where fresh water is not needed, such as in toilets or for garden use.

---

[4] Danielle Gignac and Steven D. West, "Life, Death, Water and Cape Town." https://ca.thegospelcoalition.org/article/life-death-water-cape-town/

[5] https://en.wikipedia.org/wiki/Residential_water_use_in_the_U.S._and_Canada

Agriculture accounts for a very large proportion of water consumption worldwide. It is worth asking: Are the crops grown in areas that require massive amounts of irrigation the best ones to be growing? If we allow the local climate to inform the types of plants grown, we are much more likely to work *within* the ecological capacity that a locality holds. While this may not be possible in every area, there are undoubtedly many parts of the world where resources are being over-extended in order to meet "wants" rather than "needs."

Soil may be the next most significant element pertaining to the flourishing of all life. There is significant evidence, both scientific and sociological, that the health of the soil determines the health of those who live on it, use it and consume its fruits.

It is worthwhile to investigate the scales of land use which allow for the greatest health of the soil. From a Christian perspective, we are concerned with identifying the morally right way to use soil. If fruitfulness is to be one measure of morality, perhaps we can measure our cultivation methods by how fruitful they are. However, quantity is not the only measure of fruitfulness. The analysis also ought to involve vigour and fecundity.

Systems of agriculture which are based upon cycles of soil depletion and costly artificial inputs (such as fertilizer, pesticides, herbicides, etc.) in order to efficiently work at a particular scale raise some concerns. They don't allow the soil to build the nutrients and organisms it needs for natural perennial health, but use it purely as a medium for extraction rather than as the life-provider it is. The scale is not entirely to blame, however. It is also the techniques that are used on the land that affect its fertility. There are ways of cultivating the land that enrich and build up soil fertility. Practices like crop rotation, cover crops and successive planting can contribute to building soil quality over time.

Within the soil itself is a massive ecosystem of microorganisms—it is literally pulsating with life. The food we eat obtains its nutrients from the soil it grows in. Therefore, if our food is to be nutritious, the soil must be rich in nutrients. This leads to our next topic: food production.

## 1. *Food production/agriculture*

Throughout the world today, one of the largest and most demanding uses of land by humans is for food production. We'll discuss land in more detail later, but there are a few topics under food production and consumption that are worth discussing here. Food security and food sovereignty have been topics of great interest in recent years, and Christians need to be thinking about them.

Food security "deals with the just and fair supply of food to human beings."[6] Food sovereignty is "the right of people and communities to decide and implement their agricultural and food policies and strategies for sustainable production and distribution of food."[7] Simply speaking, food security can be more of a top-down approach to ensuring the food supply is equitable, while food sovereignty often has a bottom-up approach whereby consumers exercise control and influence over their food systems. We need to think about the role of both producers and consumers in our contemporary food systems. Not every way of producing, distributing, selling and consuming food is equally ethical. There are some who criticize "food security," saying that it doesn't go far enough to ensure justice along the entire supply chain. For example, even if we can ensure food is supplied fairly to all, that doesn't necessarily mean that the food that is delivered is nutritional, culturally appropriate or sustainably grown. Consumers should try to make ethical choices whenever possible, but what can consumers do when transparency along the food chain is lacking, and the difference in cost of more ethical choices in food can be prohibitive for some? The entire burden cannot be placed on consumers. Producers have the moral responsibility to do their best to use fair and beneficial practices so that consumers are able to buy without complicity in an unethical system.

**Not every way of producing, distributing, selling and consuming food is equally ethical.**

It is worth noting that not all food is created equal. Even the same types of products can have drastically different nutritional content,

---

[6] Christian Coff, Michiel Korthals and David Barling, "Ethical Traceability and Informed Food Choice," in *Ethical Traceability and Communicating Food*, ed. Christian Coff, et al., The International Library of Environmental, Agricultural and Food Ethics 15 (New York: Springer, 2008), 9.

[7] Michael Windfuhr and Jennie Jonsén, *Food Sovereignty: Towards Democracy in Localized Food Systems* (Rugby: ITDG Publishing, 2005), 13.

depending on how they were grown. With regard to access to healthy food, there is more to the discussion than simply eating more fruits and vegetables. The *quality* of the food matters, which is dependent on how it is grown—one tomato may be more nutritious than another.

How ought Christians to view food and the environment? What about the environments that livestock are raised in? In some regions, the living conditions for animals are extremely controlled and artificially designed by people. In other areas, that's either not possible or it's deemed undesirable, and animals roam more freely. Some of these environments are healthy and others are not. "Ethically raised" animal products are becoming more popular in some parts of the world, as consumers reject conventional factory-farm methods that are considered inhumane or unhealthy. Many people are also looking for products that don't contain hormones or antibiotics, out of concern for how those will be transferred to their own bodies when consuming them.

Numerous labels are being applied today to food products. From "Fair Trade" to "organic" to "grass-fed" to "hormone-free," all of these have ethical overtones. Many of these labels also come with higher price tags, sometimes just because of the cost for certification. However, it is worth considering if these labels truly ensure more ethical practices than conventional methods. In many parts of the world, these labels are not present, but the food produced may very well achieve the same standard. In other parts of the world where there may be a greater disconnect between consumers and the food sources and details of production aren't provided, the classifications are helpful to know exactly which protocols are being followed. Sadly, there are many unethical methods applied today that meet minimal standards for food consumption by regulators, but if consumers were privy to the full information, they would likely think twice about eating the products. If one wants to guarantee that they are supporting ethical practices, for the benefit of both producers and consumers, food labels can be a helpful tool, though they are not the only way to source food ethically.

> **If we do not pay a fair price for the product, it is always the poorest labourers who suffer the most.**

We should always be willing to pay the true value for something. If we do not pay a fair price for the product, it is always the poorest labourers who suffer the most. If the only way for producers to sell cheap goods is by paying their workers extremely low wages, they are guilty of exploitation. If we have alternatives, but choose to support them by buying their overly cheap goods, we are indirectly guilty of exploitation as well. Those with financial advantages must care for the workers at every stage of production, and not merely look to buy the cheapest food available. Even for our own sakes, it is often the case that cheaper meat and produce are raised with less care and have less nutritional value than other options. Conscientious food purchases can be one way that we can care for those who produce our food.

What about genetically modified foods (i.e., GM foods or GMOs)? This process involves altering the genetic material of organisms in ways that don't occur naturally through mating or natural recombination. Sometimes this involves combining DNA from one species with the DNA from an entirely unrelated species, which means we might get animal genes with plant genes, for example.

There are various perspectives on the ethics of this method of food production. Some will say that altering the genetic makeup of species is not our role. God is the Creator and we shouldn't try to take his place. Genesis says that he made everything "according to its kind" (Genesis 1:11–12), so genetic engineering is disrupting the order that God created. Another perspective would suggest that God gives us the knowledge and ability to alter genetics, and it can have results that improve the yield of crops or create disease-resistant varieties that help human life, especially in developing nations where the reliability of food production can be more volatile.

Let's consider the context for the creation of GM foods. One of the most widely used types of GMOs are herbicide-tolerant crops. This means that herbicides can be applied broadly to crop fields to kill weeds growing up among the desired cash crops, and the desired crops are unharmed because they have been genetically engineered to withstand the herbicide. While this initially sounds like a very efficient way to promote the growth of the desired plants and remove competition, there are some downsides. "Superweeds" have been one outcome—weeds that have developed their own resistance or immunity to the herbicides now grow and reproduce. This becomes increasingly problematic as new herbicides need to be created to battle these stronger weeds. This is a similar problem to "superbugs" that become immune to medicines and are harder and harder to control. Herbicide and pesticide application also means that consumers may be exposed to these chemicals, which can in some cases be absorbed into the food. The herbicide-tolerant approach to GMOs takes the stance that we are in competition with nature, rather than finding ways to work with the systems that already exist. For example, crop rotation and polycultures (the simultaneous cultivation of several crops) are methods that can help to reduce weed proliferation without chemical inputs. GM crops, however, encourage monoculture farming. In a natural ecosystem, biodiversity emerges unprompted, as each species finds its own conditions to thrive and is often able to benefit others around it.

We don't yet have enough evidence of the long-term health effects of GMOs on the people that consume them. We do know that there are also innumerable species that are negatively affected by the pesticides and herbicides that are applied to engineered varieties. Bees are one of the most widely talked about insects affected, because bee pollination is vital for many crops to produce fruits. There have been massive bee deaths in recent years that are attributed to neonicotinoid pesticides, which will have consequences for crop growth around the world.

> **One of the moral dilemmas GMOs introduce is the patenting of genetic information.**

One of the moral dilemmas GMOs introduce is the patenting of genetic information. Large agri-businesses patent their seeds and then sell them to farmers. Because of the legal structure of the patents, the farmers are not allowed to save seeds from their crops since they belong to the corporation. Farmers are essentially forced into buying seed year after year, despite the fact the plants they grew produced viable seed. Lawsuits have already been laid against farmers who have been found with patented seed on their fields—either from cross-pollination or having them blown in from another field—despite the farmers having no intention of growing these crops. The high cost of these crops can easily lead to debt, and the inability to save seeds means this must be a yearly expense. If the growing season is not good one year, it becomes a net loss which is hard to make up the next year after purchasing seed yet again.[8]

---

[8] North American farmers have one of the highest rates of suicide among any class

"Terminator" seeds have built-in sterility, and they have been proposed as one alternative to seed-bearing GM crops. This seems intrinsically to go against God's order of creation in which *all* plants were seed-bearing. What is the long-term impact when plants are bred to be sterile and lose the ability to reproduce altogether? This concept is predicated on a system of food production that is sustained exclusively through human intervention, which is a system fraught with social, political and economic issues, to say nothing of environmental ones.

Since the environment is God's gift to us, misusing and polluting it would logically be disrespectful and dishonouring to God. There is a sensibility, at least in some nations, that we ought to take responsibility for the waste we produce, and manage it so that it does as little harm as possible. The best solution, of course, is to reduce the waste we produce in the first place.

**...it is considered more profitable for companies to create products that wear out quickly so people need to buy new ones.**

Today, much of the world has a throwaway mentality. Products are made to serve a purpose for a limited time, and then are discarded for something new. In a growth-driven economy, it is considered more profitable for companies to create products that wear out quickly so people need to buy new ones, rather than make long-lasting products or repairable ones. Planned obsolescence is a huge contributor to landfills and is a decision made by corporations who don't pay the cost of the pollution they create. This has global implications.

Where does all of that waste go? Many nations pay to export waste to other countries, often where regulations on waste management are less strict or costly and pollution is heavier. Canada, for example, exports a significant amount of its recyclables to China, so we often don't see the effects of the waste produced by our consumption. A lot of our waste ends up in ecosystems that are not equipped to handle it; for example, the "Great Pacific Garbage Patch," a massive gyre of marine debris (mostly plastics) in the central North Pacific Ocean. It

---

of workers. The reasons and causes are multifaceted, but it is clear that many farmers feel overwhelmed and are in despair.

too often means poorer nations pay in ecological health for the wasteful lifestyles of richer nations, and this is an injustice toward them.

Resource extraction is one of the topics that has many sides to it, and has less clear distinctions between right and wrong. This field includes practices such as mining, forestry, water and fossil fuel extraction and energy production (renewable or non-renewable). All of these industries produce products we use in our daily lives. If there are cases where it is unethical to obtain certain resources, is it also unethical for the end users to have the products they create? Is there a difference in ethical responsibility for producers and consumers? Wendell Berry asks, "Is there…any such thing as a Christian strip mine?"[9]

## 2. Sin affecting access to good land

Generally, ethics involves accepting certain limitations on how we act in the world. So, what are the limitations to using the land and its resources as part of our cultivation and our dwelling on the earth?

In the Bible, the relationship people have with the land is often tied to their morality. Adam and Eve were expelled from the Garden of Eden because of their sin. God gave Abraham and his descendants the gift of Canaan, "a land flowing with milk and honey" (i.e., overwhelming fruitfulness and fertility). God promised his redeemed people, the Israelites, that they would live in the fruitful land of Canaan if they obeyed his laws and walked in his ways. Deuteronomy 28 records covenant blessings and curses for either obedience or disobedience—many of them deal with the fertility of the land, fruit and flocks. The climactic covenant curse was exile from the fruitful land.

The Israelites who disobeyed and tested the Lord in the wilderness would not see the Promised Land, and would not experience the fruits and blessings that God had prepared. Rebellion and contempt meant exclusion from the blessing of good land. Suffering in the desert for forty years was their reward. Caleb, on the other hand, was rewarded for his morality by being given a special portion of the land for him and his descendants (Joshua 14:13–14).

---

[9] Wendell Berry, *The Gift of Good Land: Further Essays Cultural and Agricultural* (Berkeley: Counterpoint Press, 1981), 299. A *strip mine* is a mine where the entire top layer of soil and rock is simply stripped away, allowing access to the minerals embedded below the surface. Such a mine permanently destroys the land and habitat.

Although Christians are under the new covenant rather than the old covenant, there are practices that we participate in today which prevent us from experiencing the fruits of the land. There are direct and indirect natural consequences of our immoral choices. Jeremiah 12:4 says,

> How long will the land lie parched
>   and the grass in every field be withered?
> Because those who live in it are wicked,
>   the animals and birds have perished.
> Moreover, the people are saying,
>   "He will not see what happens to us."

May this not be true for us today as well? Obviously, there are natural cause and effect relationships: if we pour pollution into water, the water will be undrinkable; if we fill the air with toxic chemicals, the air will be toxic. But there is more than just natural cause and effects. We live in God's universe—a moral universe—and God can punish us for our hubris, folly and sin. If we sin against him in our stewardship of the earth, then he can chastise us through the medium of the natural environment.

## ENVIRONMENTAL ETHICS AND ANIMALS

How are we as Christians to relate to animals? Like the land, animals also belong to God:

> …for every animal of the forest is mine,
>   and the cattle on a thousand hills.
> I know every bird in the mountains,
>   and the insects in the fields are mine (Psalm 50:10–11).

Though not spiritual beings like humans, animals do have an awareness which other parts of creation lack. They are able to feel and to respond to their surroundings. As creatures made by God, they have value in and of themselves, and God cares for them (even though he doesn't value them as highly as human beings). The familiar passage of Matthew 10:29–31 describes how God's will pertains to the life of animals, but that humans are worth even more to him. Adam's naming of the animals established man's authority over the animals, but it also established man's relationship with animals as their caretaker (Genesis 2:19–20).

What is proper care for animals as part of our stewardship of creation? Every creature has dignity and integrity by virtue of being created as a living being by God. All living things should be permitted to exercise their normal behaviour as determined by God. We ought to aim at the positive goal of helping them flourish, rather than the negative goal of merely minimizing their suffering. Do humans have a role to ensure the protection and survival of other species? How do the results of our actions on the environment implicate us in the destruction of animal life, whether directly or indirectly? Animals function as part of an ecosystem and have purposes that contribute to and benefit from the environment in which they live.

> **Every creature has dignity and integrity by virtue of being created as a living being by God.**

Before the Fall, God's instructions indicate that plants were the only food provided for both animals and humans to eat. After the Fall, however, death entered the world, and the first recorded act of killing an animal was done by God himself to provide clothing for Adam and Eve (Genesis 3:21). After the flood, we get God's clear direction to Noah, "Everything that lives and moves about will be food for you. Just as I gave you the green plants, I now give you everything" (Genesis 9:3).

If killing and eating animals is permitted, there must be right ways and wrong ways of doing so. Intentionally inflicting unnecessary pain and suffering is wrong. Trophy hunting, for example, involves the selective hunting and killing of animals for the purpose of recreation, rather than for eating. If we look at motives, often the purpose is to bolster one's pride. On the other hand, sometimes animals are killed because they are overpopulating an area, and are likewise not necessarily used for human consumption. But the motive in this case is to enable the ecosystem to function in a balanced way, which the quantity of animals hinders. If we ask which case better demonstrates human stewardship over creation, the second case seems more tenable. There ought to be an appropriate respect for animals and an understanding of our proper relationship to them as given by God.

While hunting harvests a relatively small portion of the animal meat people eat, factory farming has spawned rigorous debate about the ethical treatment of animals. Animals are often manipulated in order to meet human-driven desires, by such means as giving them hormones or steroids, which alter their natural functioning and growth. Their living conditions are about efficiency and productivity rather than welfare. Proponents of these practices tend to view animals as consumer products. But is that how God views them? In Exodus 23:12, the Sabbath law was applied toward working animals as well, to give them rest. This indicates a concern for the well-being of animals. As well, every seven years, the fields were to be left fallow in order that the wild animals might be able to glean some food from them. Proverbs 12:10 also says that "the righteous care for the needs of their animals."

Caring for animals seems to necessitate a few things. We must respect the dignity of their being, including ensuring they are able to function according to their God-designed purposes. We also ought to be respectful and humane when we have to take their lives. This means not taking more than necessary, and not wasting anything that we do take. This would logically rule out things like killing elephants for their ivory tusks or creating living conditions on farms that are overly harmful to the health and well-being of the animals.

We will add a small note on wildlife. Though there are many animals that are not incorporated into any sort of human activity, based on the principle that all creatures belong to God, wildlife is also valuable in and of itself. As all creation does, wildlife gives glory to God. As stewards of the earth, we ought to ensure habitats are maintained for all creatures to thrive in. Our ignorance (both lack of knowledge and lack of concern) so far has caused the irretrievable loss of innumerable species. With the breadth of knowledge available today, we have no excuse to continue polluting our oceans, destroying our wetlands, overharvesting trees or otherwise ruining the sensitive ecosystems on which wildlife depends.

## ENVIRONMENTAL ETHICS AND PEOPLE

As we've already mentioned, the environment includes the humans who live within it. Human culture and society is not detached from an environmental context. Both our attitude toward and our practices in the environment are moral issues. Since we have an ethical obligation to other people, and an ultimate ethical obligation to God, we have objective moral responsibilities for environmental care. Fulfilling these moral duties brings about positive consequential benefits. Ethical behaviour in and for God's creation—our common environment and home—directly and indirectly generates and sustains the type of healthy, functioning environment necessary for human flourishing.

## 1. Population growth and density

Population growth and density is one of the most discussed topics in relation to environmental sustainability. God gave humans the mandate, "Be fruitful and increase in number; fill the earth and subdue it" (Genesis 1:28). This does not address the environmental impacts of the imperative, so we are left to fill in the blanks. Is there a limit to the number of people a given environment can support? Where do we get this information? Are there natural systems that balance out population (eg. famine, drought, disease), and are we to accept their function as levelling mechanisms? (Different societies seem to have conflicting ideas about this. Some societies acknowledge that a large population is a burden on their resources and space, so they limit the number of children a family can have. Others encourage large families—for example, in some rural contexts, more children mean lots of help to work the land and sustain the community.)

As of November 2018, the population of the world had reached 7.7 billion. It is expected to reached 9 or 10 billion by midcentury.

One of the biggest topics related to world populations is whether the earth can produce enough food to feed its quickly growing population. If current production falls short, can different methods be used to safely and sustainably raise the levels of production? Since different agricultural models may entail different environmental complications, each alternative needs to be thought through with sufficient care. Clearly, however, a booming population requires adequate housing, sanitation, food, water, etc. There are real debates about how best to accommodate these realities. When it comes to housing, for example, some believe that urban high density is better for the environment because it limits the amount of land that needs to be developed. Yet, many cities experience urban sprawl, where good land for food production is made useless for those purposes. If global food production really is a concern, should we prioritize keeping prime agricultural land for food rather than housing? It is entirely possible that the world can sustain a larger population, but perhaps not with our current lifestyles. In order for more people to live, some who overconsume may have to consume less.

Every ecological region on earth has a particular carrying capacity—the number of living organisms that can be supported by the resources of an area without environmental degradation. This can be hard to define exactly in a given region, but it is clear when the capacity is reached—available resources cannot keep up with the needs of a population. Urban areas around the world simply cannot be sustained by their own resources. They must draw on surrounding areas for food, water, power, etc. It would seem, however, if cities need to do this, they must also ensure that the areas they are drawing from have enough to sustain themselves. To take so much from their neighbours that it leaves them destitute is harmful.

**Simply consuming less, and producing less waste, has real and beneficial effects.**

It is worth noting that we can also harm future generations if we leave land without the ability to regenerate—if we empty aquifers and deplete soil. We should only take what we need. "If you enter your neighbor's vineyard, you may eat all the grapes you want, but do not put any in your basket. If you enter your neighbor's grainfield, you may pick kernels with your hands, but you must not put a sickle to their standing grain" (Deuteronomy 23:24–25). In the case of affluent nations, this means we must consider altering our lifestyles to ease the burden. We can debate whether this is primarily a personal or a political responsibility, but we must at least address it in our own lives. Simply consuming less, and producing less waste, has real and beneficial effects. Around the world there are some who cannot consume even their basic needs, while there are others who consume more than they should. Those in the latter category need to amend their practices.

There is also the consideration of *how* land is used. Some practices are qualitatively better than others for long-term sustainability. Usually, this requires a careful understanding of a particular landscape, and the methods that are used in one place must necessarily be altered for others with different conditions. Solutions that are easy and perfect rarely show up in land-use practices. More often than not, it requires patience and perseverance. Also, not all land is suitable for food production or pasturing animals. We must take special care of the land that is, as it is a gift from God, both for our use and also for the use of future generations to meet their needs.

We have charted the extents of the earth enough by now to know there *are* physical limits. There is no new frontier where unspoiled nature can support a population that will just continue to expand forever. We know we have limited resources, and we must determine how best to use them both for our good and for that of future generations.

## 2. Living and working conditions

Living and working conditions in many countries are partly a result of environmental factors. However, the work itself can also create environmental problems which further degrade quality of life. For example, pesticides that are used in commercial plantations can be harmful for the workers. Or, inhabited areas can be destroyed by the creation of dams or deforestation. In other places, pollution generated by one group is harmful for another group.

Which nation's workplace standards are applied on farms where the products are intended to be shipped internationally? Are we buying products produced by processes that caused harm to the workers— where they are not protected against toxic chemicals or other dangerous practices? In our global economy, the disconnection between producer and consumer often means that consumers rarely have any concept of the living conditions of the producers. Health and safety standards can vary widely throughout the world (eg. First World consumers vs. Third World producers). The longer the supply chain, the harder it is to keep track of what occurs at each stage.

There are ethical concerns with regards to whose standards apply with multinational corporations. Lower compensation under other jurisdictions can allow companies to obtain a greater profit without having to contribute to the well-being of the workers. If it is a company based in a wealthy nation getting products produced cheaply in a poorer country with less thorough workplace regulations and safety protocols, should the company

*If we are to care for our neighbours around the world... consider how our everyday purchases may assist in improving their lives.*

have to operate according to the rules of their own nation, and thereby actually improve the conditions that people are working in relative to other local options they have? Many companies will offer slightly improved workplaces from what locals are used to, but are unwilling to apply their home nation's standards. Their whole reason for operating internationally is for cheap labour. If Christians are in decision-making positions in these companies, is it morally satisfactory to merely bring people a step higher while receiving great profits oneself? Thankfully, consumers do have choices, though it often involves financial consideration. Designations like "Fair Trade" have been developed to ensure people are paid fairly for their work in the global marketplace. Farming and artisan co-operatives have been started to give control and profit directly to the workers. If we are to care for our neighbours around the world, it would do us well to consider how our everyday purchases may assist in improving their lives.

### *3. Technology*

Technology has helped us to overcome many of the effects of sin in our world, such as healing and preventing disease, providing global aid, getting the gospel to unreached people groups, etc. But it has also enabled those with knowledge and resources to exercise dominance over others and over creation. When it comes to non-human entities like the environment, what are the structures needed to set limits to power and create rational self-restraint?

In many ways, technology has allowed us to ignore the limitations of nature and to impose upon it our concept of limitless growth. If we can solve the problems related to *how* we can achieve something, we often stop asking, "Why would we?" and "Should we?" We have circumvented natural boundaries by just putting more technological effort into it. If a farmer's field isn't productive enough (perhaps because it's being overburdened), the conventional solution is to just add more chemical fertilizers and not worry about all of the other problems that depleted soil signify, such as loss of microorganisms, nutrient depletion, etc.

It is also quite clear that natural systems possess resilience and adaptability, in comparison to human-made ones. For example, large coastal mangrove forests mitigate tidal destruction, erosion and siltation of waterways. They require no human maintenance, are living systems that reproduce on their own, and they provide habitat for other creatures. Human attempts to create systems with the same function are extremely costly, require extensive engineering and long-term maintenance. The mangroves are so elegantly and holistically simple compared to our efforts.

The frameworks that technological products create impose lifestyles based upon the interests of those who produce them. Here in Canada, building dikes in northern First Nation communities to protect them against flooding implicitly signifies a lifestyle of defiance toward the natural environment and a preference for stasis. Many would argue these interventions are contrary to the way of life of the culture they're intended to serve. Systems of housing and the associated technologies have also produced the type of First Nation reserve communities we see today, with gridded streets, boxy houses and infrastructure not suited to the climate or economic capability of the communities.

It would be far more sustainable to promote ethical cultural paradigms in relation to the environment, and use technology as tools to achieve that. Currently we are more often working the other way around, and seem to be at the mercy of what technology produces. We are unconsciously determining our fate and seeing it as a course of action, not as a choice.

### 4. Culture

It is important to be attentive to local cultures when addressing environmental issues. Consumerism can have a levelling effect on cultures, and create a world of assumed sameness, when in fact cultures have been intrinsically shaped by their surrounding environment. Cultures have developed through the particular ways we have learned to mediate between ourselves and the rest of creation around us. Specific local environments, known intimately by particular cultures, are where virtues can be put into practice. People are more likely to care about a place that has cultural meaning to them.

***People are more likely to care about a place that has cultural meaning to them.***

It is also important to understand the underlying assumptions that cultures, including one's own, make toward their environment. While some societies get by with outsourcing all of their damage (eg. plastic and electronic waste shipped to other countries), others that are more directly land-dependent need the integrity of their environment maintained in order to continue their culture and care for the land properly. There are communities around the world who are being forced off their land so that multinational corporations can use it. Rainforests are being cleared for plantations or herding; rivers are being dammed, flooding indigenous villages. While those who care for and live on the land are in a position to see environmental details that are vital for their culture and their livelihood, those who merely use the land for extraction are imposing their own clashing cultural framework. We agree with Wendell Berry, "The more local and settled the culture, the better it stays put, the less the damage. It is the foreigner whose road of excess leads to a desert.... But a man with a machine and inadequate culture...is a pestilence."[10]

> **Expectations for quality of life must also be understood from within a culture and not imposed from outside.**

Expectations for quality of life must also be understood from within a culture and not imposed from outside. It would just not be possible for everyone to maintain a North American standard of living—with the consumption and waste habits that are part of it—on a global scale. It is probably worth questioning if this is a good standard of living to have at all. It is better to take principles of well-being and address them within culturally relevant means. For example, if human waste management is an issue in a rural African community, it would not make sense to install a sewage lagoon with flush toilets in every home. Instead, decentralized compost toilets and education on how to manage them over time would be a more appropriate solution. It can also be damaging even in mission settings, when outsiders set up camp with all of the comforts of their home nation in the midst of those who have far less. This can create a setting of covetousness, as well as set up unrealistic ideals that locals then strive to attain. For example, if a Westerner has a metal roof on their house, it can be a status symbol for local people to save up enough money to buy one. This can take precedence over other needs in their life and be an unnecessary pursuit and use of their money.

---

[10] Wendell Berry, *What Are People For?* (Berkeley: Counterpoint Press, 1981), 8.

When Western nations send foreign aid to Third World countries, it often comes with many expectations of the type of aid that is needed based on Western values, as well as significant amounts of packaging. Many of these countries do not have the sophisticated infrastructure to manage the amount and type of waste that is imported, which then ends up causing further pollution and all of the problems that come with it, including contaminated drinking water. It is the responsibility of the nations sending aid to consider the cultural framework of those they are helping and to act thoughtfully.

### *5. Economy*

True economy is not merely "maximization of profits." It is careful management of available resources, which requires restraint and an understanding of limits imposed by reality. Since the earth is the Lord's, we are all using his resources, which are part of a common trust, provided to us for our common good. When a profit-based economy drives resource use, careful management and the common good get sidelined. As personal motives conflict, social unrest can also result. We have seen this time and again around the world. Mismanagement of resources can lead to conditions like drought or famine. When faced with few alternatives for survival, acute physical need can drive sinful humanity to acts of injustice, greed and violence. Desperate people can do desperate things.

We also face the issue of absentee landlords in many parts of the world. Multinational corporations can extract resources from an area, causing significant local ecological damage, while the rich executives may live in other places and never experience this degradation.

> **We must live with love for our neighbour, and develop a culture of care and stewardship of the common ground on which we all depend.**

How can one love one's neighbour when they are literally flooding them out of house and home with a dam reservoir? The economic and social costs of using shared environmental resources must be borne by those who use them, not by those who happen to be in the way. Proper economics brings into the equation the life of all of those impacted by environmental actions.

Proper economy of means and resources would look at the entire system to see how best to manage things. We ought to let the grace we have received, in both the gift of the natural world and the gift of redemption, be evident in how we extend grace to others and to creation. Christian spirituality encourages moderation and the capacity to be content. We must live with love for our neighbour, and develop a culture of care and stewardship of the common ground on which we all depend. A functioning economy that is environmentally sensitive would help ensure this.

## CONCLUSION

Though often left out of discussions in Christian ethics, we have a clear biblical mandate to care for the world we have been given by God. We have a positive Christian ethic to care for creation. We would also argue that practising environmental ethics is good for us and helps produce virtuous character. When we thoughtfully consider the world around us, we work toward creating a harmony between our inner and outer worlds. We develop our spiritual life as well as our physical life in the world. If we don't exercise care toward the *world* that sustains us, how will we develop real care toward the people it sustains? If we are to love our neighbour, we cannot use ignorance as an excuse.

**If we don't exercise care toward the world that sustains us, how will we develop real care toward the people it sustains?**

Any practise of morality will take effort and self-control, and affect our daily decisions. In considering the world around us and truly seeing it as belonging to the Lord, we are obligated to examine our lifestyles and attitudes toward the environment. To live as good stewards of the good earth God has given to us is essential to a virtuous life.

As we have seen, there are multiple reasons why Christians should care for God's world. One special reason is that Christians love Christ, and Christ is the central focus of the created order. He is also its maker and sustainer.

In the beginning was the Word, and the Word was with God, and the Word was God. He was with God in the beginning. Through him all things were made; without him nothing was made that has been made (John 1:1–3).

The Son is the image of the invisible God, the firstborn over all creation. For in him all things were created: things in heaven and on earth, visible and invisible, whether thrones or powers or rulers or authorities; all things have been created through him and for him. He is before all things, and in him all things hold together (Colossians 1:15–17).

In the past God spoke to our ancestors through the prophets at many times and in various ways, but in these last days he has spoken to us by his Son, whom he appointed heir of all things, and through whom also he made the universe. The Son is the radiance of God's glory and the exact representation of his being, sustaining all things by his powerful word (Hebrews 1:1–3).

The fact that Christ made the world and sustains it, proves that it is valuable to him. As Christ's followers, we must value what he values. If nothing else moves us with concern for the environment, surely love for Christ must compel us. He did not need to create anything, yet he chose to create a physical environment to be our home. The world is not ours; it belongs to Christ. It is our conviction that Christ wants his property to be well-managed and cared for by his stewards.

**As Christ's followers, we must value what he values.**

Not only is Christ the maker and sustainer of the world, he is also its liberator and redeemer. Although creation groans in bondage, one day it will enter into freedom and organic wholeness because it will share in the redemption of the children of God. Paul writes:

> I consider that our present sufferings are not worth comparing with the glory that will be revealed in us. For the creation waits in eager expectation for the children of God to be revealed. For the creation was subjected to frustration, not by its own choice, but by the will of the one who subjected it, in hope that the creation itself will be liberated from its bondage to decay and

brought into the freedom and glory of the children of God (Romans 8:18–21)

The eschatological vision in Revelation is of a glorified physical and spiritual order. Our eternal home is not in an ethereal spirit-realm—it is a material environment.

Perhaps nothing proves the goodness of the physical realm more than the resurrection of Jesus Christ. He was *bodily* raised from the dead. Christ is embodied for eternity. We shall be as well. Our future is to be clothed in glorified bodies, living in a glorified physical universe, which is the home of righteousness. All spiritual, moral, mental, emotional and physical effects of the curse and sin will be removed forever. There will never be environmental damage, exploitation or harm. We will cultivate the inherent properties of the new heavens and new earth, all to the glory of God. The environment will be more aesthetically beautiful, more fruitful and more satisfying than we can imagine. It will be eternally sustainable. When we are there, in that environment, we will find both work and rest. But more than anything, it will be our Creator and Redeemer who will give us joy, since we will live in his presence. Then, for the very first time, we will begin to truly comprehend what it means that the righteous will flourish.

## REFLECTION QUESTIONS

1. What is your culture's view of the environment?

2. Is taking care of the environment an important issue in your church? Should it be taken more seriously?

3. Martin Luther is widely reputed to have said, "Even if I knew that tomorrow the world would go to pieces, I would still plant my apple tree." What does this mean? Do you agree with this sentiment?

4. Are there areas in your life which should be altered to reduce unnecessary impact on the environment?

5. What other biblical passages speak to issues of environmental care, stewardship and proper human dominion?

# 10

# Ethics of church leadership

There is no area of life that is outside the realm of Christian ethics, because every area of life is judged by the standard of Christ's absolute goodness and moral authority. As a result, biblical and Christian ethics are concerned with some topics that are of little or no interest to secular ethicists and moral philosophers. The ethics of church leadership is one such issue. The world may notice when scandals take place among prominent church leaders, but the daily attitudes and moral responsibilities that church leaders are to have are overlooked and ignored. It is very important that we understand what Christ-honouring and Christ-imitating church leadership looks like. Spiritually healthy leaders provide a positive example for others to follow, and this means that healthy and godly leaders can help other believers grow in holiness and spiritual maturity.

Unfortunately, one of the biggest practical problems facing churches around the world today is that many church leaders have a very unhealthy, unbiblical and unethical view of their own authority, power, position and role in the church. This phenomenon transcends cultures, languages, ethnicities and geography. Abusive church leaders exist in both large and small congregations. They are found among the poor and the rich. They may work in churches where people are educated or uneducated. Cultural factors often play a role. In the Western world, some pastors treat their church like a corporation, and they act like they are the company president. They can be arrogant, demanding and impatient. In other places, pastors demand that they be given far too much social power, honour and respect. Some pastors insist on being treated like chiefs, with the members of the church serving them. All of these attitudes are wrong, and because of the influential role that leaders have in the church, they can have far-reaching negative consequences.

> *... many church leaders have a very unhealthy, unbiblical and unethical view of their own authority, power, position and role in the church.*

To cultivate proper church leadership, we must do our best to allow the Bible's instructions to overrule our culture's expectations and our own sinful hearts. There may be some flexibility for how church leadership is expressed in different cultures, but these cultural forms must not violate the basic principles for leadership that are laid down in Scripture. Given the fact that the structure of church government is a hotly debated theological topic—and has been throughout church history—we will not focus on the proper structure for church government. Some denominations have authoritative presbyteries and synods; others are made up of independent churches that only recognize local church leadership. We will not discuss whether or not bishops have more authority than pastors, or if one pastor in a congregation should have more authority than the other elders. What we will discuss, however, are the moral attitudes that ought to characterize *all* of the leaders that Christ has given to his church. The best way to do this is to turn to God's Word.

## SERVANT LEADERSHIP: SELECTED BIBLE TEXTS

### 1. Numbers 12:3

(Now Moses was a very humble man, more humble than anyone else on the face of the earth.)

This parenthetical statement describes Israel's first great leader. Moses was a strong leader, but he was characterized by radical humility. You will recall from Chapter 3 that being "completely humble" (Ephesians 4:2) is the first mark of living a life worthy of the calling that we have received. Leaders cannot be godly if they forsake basic Christian virtues. In fact, this is one of the most basic principles of all: church leaders need to be consistent examples of godly living. If Christians are to be humble, then of course Christian *leaders* are to be humble! There is not a special code of morality for church leaders, nor are they exempted from the moral requirements every believer is under. However, they need to excel in morality so those they are leading have an example to follow. We will have more to say about this later.

> **Moses was a strong leader, but he was characterized by radical humility.**

### 2. Deuteronomy 17:14-20

When you enter the land the Lord your God is giving you and have taken possession of it and settled in it, and you say, "Let us set a king over us like all the nations around us," be sure to appoint over you a king the Lord your God chooses. He must be from among your fellow Israelites. Do not place a foreigner over you, one who is not an Israelite. The king, moreover, must not acquire great numbers of horses for himself or make the people return to Egypt to get more of them, for the Lord has told you, "You are not to go back that way again." He must not take many wives, or his heart will be led astray. He must not accumulate large amounts of silver and gold.

When he takes the throne of his kingdom, he is to write for himself on a scroll a copy of this law, taken from that of the

Levitical priests. It is to be with him, and he is to read it all the days of his life so that he may learn to revere the Lord his God and follow carefully all the words of this law and these decrees and not consider himself better than his fellow Israelites and turn from the law to the right or to the left. Then he and his descendants will reign a long time over his kingdom in Israel.

The nation of Israel is not identical to the church, and pastors are not kings! Nevertheless, there are principles in this text which can be helpfully applied to church leadership.

First, the king needed to be part of the covenant community. Church leaders, therefore, need to be part of Christ's body—they must be born again. Second, the king was not to acquire many horses. This stipulation is tied to military technology. The king was to trust God, not his army and cavalry. Third, the king was not to take many wives. This was intended to speak more about entering political alliances than about the morality of polygamy. Fourth, the king was not to amass too much personal wealth. Although it is not a direct application, church leaders must be careful that they do not put too much stock in numbers and the size of their flock. They must not seek out alliances that will cause them to trust others more than God. And, very importantly, they must not try to grow rich at the expense of their people, or make money their idol. Fifth, the king was to be a person who was intimately acquainted with God's law. Church leaders need to be people who know and love the Bible. Sixth, the king must not "consider himself better than his fellow Israelites." This is a massively important principle for leadership. Church leaders must not think that they are any better than anyone else in the church. In fact, they should recognize that they stand only by God's grace and gifts.

Being appointed and set apart as a church leader should be deeply humbling. If a leader believes or acts like they are better than others, they are experiencing a real moral failure.

### 3. *Ezekiel 34:1–31*

This passage is too long to reproduce here, but it should be looked up and read carefully. It clearly reveals God's attitude toward leaders who serve themselves rather than serve his flock. God rebukes them as worthless shepherds who are under his judgement, but then he also shifts the metaphor and refers to them as bullying animals in his flock. This is extremely important for pastors to understand. To be a pastor, you must first be one of Christ's *sheep*. If you think that being a pastor means you are smarter and wiser than the "dumb" sheep of Christ's flock, you are terribly mistaken. Christ's sheep are never referred to as stupid. Christ cares for, loves and exercises authority over his flock. One image for pastoral leadership is that of a shepherd; but the human shepherd is always a sheep as well, no less than any other believer. In this passage from Ezekiel, God says that he will judge between one sheep and another. Church leaders, you will be judged by God for how you treat his sheep, and you will find out that he is your shepherd as much as theirs. In John 10, Jesus teaches that he has *one flock* for which he dies. If you are not one of Jesus' sheep, then you are not saved. But if you are bought by the blood of your Good Shepherd, how can you harass and bully others in the flock? You are no better than they are. Your attitude and actions should reflect this truth.

> **If you are bought by the blood of your Good Shepherd, how can you harass and bully others in the flock?**

### 4. Mark 10:42–45

> Jesus called them together and said, "You know that those who are regarded as rulers of the Gentiles lord it over them, and their high officials exercise authority over them. Not so with you. Instead, whoever wants to become great among you must be your servant, and whoever wants to be first must be slave of all. For even the Son of Man did not come to be served, but to serve, and to give his life as a ransom for many."

In context, James and John have asked Jesus for special places of honour and glory in God's kingdom. Jesus corrected them, then went on to teach his disciples about godly leadership. He contrasts the way the world looks at leadership with his own approach and mission. Yes, the world loves to be in charge, and the world loves to lord power and prestige over others. Jesus flatly rejects this attitude, and he denies its validity for his disciples. Church leadership is not about power and position, *it is about sacrificial service.* Christian leaders are to be *servant leaders.* After all, it is impossible to imitate Christ without being a servant. Jesus commands his disciples to be slaves of all. God the Son incarnate—the Lord Jesus Christ—did not make his disciples serve him. He served them, even to the point of shedding his blood to ransom their souls. If the Lord of the universe serves us this way, how ought we to serve others? Too many church leaders play games with words, saying that they are serving others when, in reality, they are domineering toward their churches. Church leaders must stop trying to be served, and start imitating Christ.

### 5. John 13:1-17

In this passage, Jesus wraps himself in a towel, takes the most menial job of the lowliest household slave, and washes the feet of his disciples. After performing this task—which was utterly humiliating in that culture—Jesus taught his disciples to imitate his actions. Verses 12–17 say:

> When he had finished washing their feet, he put on his clothes and returned to his place. "Do you understand what I have done for you?" he asked them. "You call me 'Teacher' and 'Lord,' and rightly so, for that is what I am. Now that I, your Lord and Teacher, have washed your feet, you also should wash one another's feet. I have set you an example that you should do as I have done for you. Very truly I tell you, no servant is greater than his master, nor is a messenger greater than the one who sent him. Now that you know these things, you will be blessed if you do them."

**If we fail to act like servants, we are actually acting like we are more important than Jesus.**

At best, church leaders are good teachers and shepherds, but they are never the church's Lord. Here, the church's Lord and Teacher placed himself in the position of the lowest slave. Church leaders: Do your churches see you as someone who acts like this? (In an authentic way, not hypocritically; wanting to genuinely serve, not just wanting to appear humble or godly.) Notice that Jesus said that he has set us an example which we are to imitate. If we fail to act like servants, we are actually acting like we are more important than Jesus. If Jesus would serve in this way, who are we to decide that we're too good to engage in menial acts of service? Such humble service is the path to blessing. A church leader who lords it over others while refusing to be a servant of all is not Christ-like, and therefore is unqualified for leadership in the church. Jesus was a servant leader, and he is our perfect model and standard.

### 6. Philippians 2:5-11

> In your relationships with one another, have the same mindset as Christ Jesus:
>
> Who, being in very nature God,
>   did not consider equality with God something to be used to his
>     own advantage;
> rather, he made himself nothing
>   by taking the very nature of a servant,
>   being made in human likeness.
> And being found in appearance as a man,
>   he humbled himself
>   by becoming obedient to death—
>     even death on a cross!
> Therefore God exalted him to the highest place
>   and gave him the name that is above every name,
> that at the name of Jesus every knee should bow,
>   in heaven and on earth and under the earth,
> and every tongue acknowledge that Jesus Christ is Lord,
>   to the glory of God the Father.

The famous Christ-hymn in verses 6–11 is introduced by Paul laying out the ethical command that we must have the same "mindset" as Christ Jesus. The passage makes clear that the mind of Christ led him to humbly sacrifice himself for others. This command to imitate Christ is for every believer, which of course includes church leaders. Paul points us to the fact that the mindset of Christ is what led the Son to become incarnate. Even though he was God, the Son willingly clothed himself in humanity, emptying himself and becoming the servant of sinners. He humbled himself, not just by coming to earth, but by dying on the cross. We cannot imagine a greater act of service—in fact, a greater act of service is logically impossible. The hymn goes on to celebrate the success of Christ's atoning death, and his vindication in his resurrection and ascension. But the ethical application of the text is found in imitating Christ's humility and service. A church leader's relationship with other members of the church is to be characterized by this self-sacrificial stance of Christ Jesus.

Clearly, having such an attitude cannot be separated from practical living. If a church leader is like Christ, their church will know it. A church leader who is arrogant and refuses to serve is failing to imitate the Lord of the church. Once again, the phrase "servant leader" seems most appropriate.

### 7. 1 Timothy 3:1-7

> Here is a trustworthy saying: Whoever aspires to be an overseer desires a noble task. Now the overseer is to be above reproach, faithful to his wife, temperate, self-controlled, respectable, hospitable, able to teach, not given to drunkenness, not violent but gentle, not quarrelsome, not a lover of money. He must manage his own family well and see that his children obey him, and he must do so in a manner worthy of full respect. (If anyone does not know how to manage his own family, how can he take care of God's church?) He must not be a recent convert, or he may become conceited and fall under the same judgment as the devil. He must also have a good reputation with outsiders, so that he will not fall into disgrace and into the devil's trap.

This passage lays out the qualifications for an elder in more detail than any other biblical text. It begins by saying that it is not wrong to aspire to be a church leader, and that being an overseer is a noble and good thing. In our judgement, the New Testament words "overseer," "elder" and "pastor" all refer to the same position, but this is not crucial for the issue of the ethics of servant leadership. Notice that this passage does not set out a list of *tasks* that the elder is supposed to perform. It is not a job description. Rather, this text describes the necessary moral *character* for a church leader. Very generally, they are to be above reproach. In other words, there should not be a glaring and obvious fault in a church leader that disqualifies them from their position. No one is perfect, and mistakes will be made, even by leaders. Not only will church leaders err, they will also sin. Nevertheless, there is to be a general consistency and integrity in a leader's life and walk with God. His life in the church, home and society are all important and cannot be divorced from each other; they are all relevant to his character.

When we read through this characterization of the overseer's spiritual life, we can't help but notice that these virtues are also to be true of every believer, whether they are in a position of leadership or not. Elders are not be drunkards, but neither is anyone else in the church! Likewise, no Christian is to be violent or a lover of money. The gift of teaching is one requirement for being an elder—and obviously not every Christian has that particular gift—but all of the virtues listed in this passage are to be found in every believer.

In terms of servant leadership, there are a few essential points. The leader must be *gentle*. He must not be someone who is quick to fight and quarrel. This means he must be someone who listens and considers his own words before he speaks. He must be patient with people who disagree with him. A godly church leader does not push his people too hard; they may need a gentle nudge now and again to flourish and be productive, but this is done with a loving hand. Serious sin may require serious treatment, but the elder never wants to be overly harsh. **Domineering, authoritarian and dictatorial approaches to church leadership are completely foreign to the spirit of this text.** Pride and arrogance actually disqualify someone from being an overseer.[1] People who work on projects with a

---

[1] Notice that the rationale behind 1 Timothy 3:6 is that a new convert who is appointed to leadership may become conceited, and thus be ruined. Arrogance and conceit are deadly in leaders. It takes spiritual maturity to recognize the difference between humbly exercising authority versus operating with arrogant authoritarianism. This discernment takes time to develop. Pride hurts everyone in the church, including the conceited individual. Young leaders must take this to heart, and mature congregations must help protect those who seek leadership. They must be ready—being put into leadership too quickly will harm, not help, a potential candidate.

church leader should find that they are treated generously with grace, patience and kindness. This does not mean that a church leader doesn't lead, but he is sensitive to others. After all, since he is a follower of Christ, his goal ought to be to bless people with humble and loving service.

### 8. 1 Peter 5:1-4

> To the elders among you, I appeal as a fellow elder and a witness of Christ's sufferings who also will share in the glory to be revealed: Be shepherds of God's flock that is under your care, watching over them—not because you must, but because you are willing, as God wants you to be; not pursuing dishonest gain, but eager to serve; not lording it over those entrusted to you, but being examples to the flock. And when the Chief Shepherd appears, you will receive the crown of glory that will never fade away.

Peter's words here are simple, profound and beautiful. Even though he is an apostle of Christ with authority over these elders, he puts himself on their level. He is an apostle, but he is also a shepherd, just like they are. Peter tells the elders that they are to take care of God's flock, watching over it. There is a diligence and care here that must not be neglected. After that general exhortation, Peter presents three sets of contrasts that are to characterize a shepherd of God's flock.

### (i) Church leaders should not serve grudgingly.

They should not complain about the tasks that God has called them to perform. They must not resent the sheep. On the contrary, they ought to be willing to serve, which is how God wants them to be. Servant leadership is not achieved simply by performing menial tasks—our acts of service must flow out of a heart of love. If a church leader does not want to honour Christ and bless Christ's flock, then they cannot please God no matter how many feet they wash or people they give up their time and resources for.

### (ii) They must not be in the ministry for themselves.

It is possible to secure some dishonest gain by taking advantage of members of the congregation, or even by stealing or embezzling church donations and funds. Just as the king of Israel was not to acquire too much silver and gold, so the church leader must be content with what God provides for them through legitimate channels.

Rather than being self-interested, leaders are to be eager to serve. Pay attention to that clause! Leaders must not only be servants, they must *want* to be servants. They must be *eager* to serve. Rather than sitting back and expecting people to serve them, real Christian leaders want to pour into others and help build them up, according to their needs. A godly Christian leader gets great joy from seeing others succeed and realize their potential. Like Christ, a godly leader is eager to serve rather than be served (Matthew 20:28), and they know it is more blessed to give than to receive (Acts 20:35).

> **Rather than being self-interested, leaders are to be eager to serve.**

*(iii) The elder is not to be someone who lords their position and authority over others.*
We have already seen this. Jesus, Paul and Peter all take time to point out how important it is for leaders to serve. This repetition should help us grasp how significant it is for church leaders not to be people who enjoy status or power. There is hardly anything more opposed to Christ's example than the spectacle of an arrogant, self-serving pastor. Rather than acting like the world—where Jesus says people love to exercise power and lord it over others—Christian leaders are to be examples to the flock. Since Christ is the great example, this entails that church leaders must imitate Christ. We have already seen the example that Christ has provided for us all, whether we are leaders or not. His example is one of humility, service, self-sacrifice and love. Christ is the epitome of a servant leader. Peter says that an elder must be an example to the flock, and this requires imitating Christ. Any elder who is not a servant leader is not really an elder—they are an imposter; a wolf in sheep's clothing.

## CHURCH DISCIPLINE

Enough has been said to prove that imitating Christ's servant leadership is at the *heart* of genuine church leadership. Pastors and elders are to demonstrate godly virtues, imitate Christ and serve others in love. None of these truths, however, contradict the fact that leaders are to provide leadership. Moses, Jesus, Peter and Paul were all strong leaders. Church leaders are to be honoured and respected by the congregation (1 Timothy 5:17; Hebrews 13:17). Of course, church leaders are to earn that respect and honour, not just demand it.

Although there are a variety of tasks that church leaders need to perform, one that can be terribly difficult for both the leader and the church is the exercise of church discipline. Pastors are often put in positions of knowing sensitive information, and they are often looked to for wisdom when it comes to confronting and correcting sin. Dealing with people in crisis situations generated because of sinful behaviour and attitudes is very trying. This is where leaders must walk with the Spirit and demonstrate godly maturity. Being a humble, loving servant leader does not exclude disciplining the erring when discipline is required. But even in these extreme cases, bitterness, anger and selfishness are unacceptable. As we will see, Christ-like virtue is always required.

### 1. Matthew 18:15–20

> [Jesus said,] "If your brother or sister sins, go and point out their fault, just between the two of you. If they listen to you, you have won them over. But if they will not listen, take one or two others along, so that 'every matter may be established by the testimony of two or three witnesses.' If they still refuse to listen, tell it to the church; and if they refuse to listen even to the church, treat them as you would a pagan or a tax collector.
>
> "Truly I tell you, whatever you bind on earth will be bound in heaven, and whatever you loose on earth will be loosed in heaven.
>
> "Again, truly I tell you that if two of you on earth agree about anything they ask for, it will be done for them by my Father in heaven. For where two or three gather in my name, there am I with them."

In this text, Jesus lays out a progression for confronting and correcting someone who is unrepentant in sin. The first step is to keep the matter as private as it can be. An initial conversation should be between just two people, if at all possible. Now, if the sin was committed publicly or it is particularly flagrant (or illegal), then more people will need to be involved from the beginning. Ideally, however, Jesus envisions sin being dealt with in a quiet manner, where two people talk and the sinner repents. If they do not repent, however, then the individual who talked with them originally is to bring along a small number of witnesses. The hope is that this small group may be able to help the individual see how serious the matter is, and help them acknowledge their sin and turn away from it. If even this strategy fails to produce repentance, however, then the sin is exposed to the church. This is the last—and never hoped for—step in the process of church discipline. Even at this point, however, the goal is *reconciliation* and *repentance*. When an unrepentant sinner won't listen to the entire church, there is little else that can be done. If the congregation tells us that we are sinning and need to repent, we ought to search our hearts very carefully and be quick to listen. Where this last step is taken, and still there is no repentance, then the church must treat the person as a pagan or tax collector. How were such people treated? They were not treated as part of the church, which meant that they must be regarded as unsaved. But, of course, the church was trying to win the lost. So the goal is not punishment—the goal is salvation.

> **Steps in church discipline**
> 1. **One by one:** Point out their sin just between the two of you.
> 2. **Small group:** Go with one or two others to plead for repentance.
> 3. **The church:** If the sinner still refuses to listen, tell the church and treat them as an unbeliever, seeking their salvation.

### *2. 1 Corinthians 5:1–13*

In this passage, Paul is dealing with the failure of the Corinthian church to exercise discipline over an egregious display of sin. The Corinthians thought that they were being enlightened and spiritual by tolerating this man's sin, but Paul insists that the individual needs to be put out of the church. His language is very strong: "So when you are assembled and I am with you in spirit, and the power of our Lord Jesus is present, hand this man over to Satan for the destruction of the flesh, so that his spirit may be saved on the day of the Lord (1 Corinthians 15:4–5). Although interpreters debate exactly what Paul means by "hand this man over to Satan," it seems clear that being removed from the protection of the church is supposed to get this man to realize how far he has fallen into sin. When he is removed from the presence of the godly, he will discover just what is in his heart and soul. The goal of this is not to punish and torment him. On the contrary, it is to bring about the salvation of his soul. Reconciliation and redemption must always be the aims of biblical church discipline.

### 3. Galatians 6:1

> Brothers and sisters, if someone is caught in a sin, you who live by the Spirit should restore that person gently. But watch yourselves, or you also may be tempted.

This verse clearly affirms the necessity of having a right *attitude* when it comes to helping a person who is trapped in sin. Once again, the order of the day is for gentleness. Someone who is caught in sin does not need to be abused or attacked—they need release and healing. A godly Christian leader exercises virtue and follows the Spirit even in the most difficult of times. Humility is essential, since a proud leader is more susceptible to temptation and falling into sin. How many times have we read in these texts that Christian leaders must be gentle? A virtuous character is absolutely necessary for navigating the pitfalls church leaders will inevitably face.

*A godly Christian leader exercises virtue and follows the Spirit even in the most difficult of times.*

## CONCLUSION

In this chapter we have explored some of what the Bible has to say concerning godly servant leadership. Time and again we have seen the need to imitate Christ, to be humble, gentle, loving and self-sacrificial. It is scandalous how many pastors and church leaders demonstrate the opposite tendencies and characteristics in their ministries. We cannot even imagine the amount of damage done in churches around the world through the failure of pastors and leaders to live out a Christ-like servant leadership. Far too many church leaders are forceful, proud and aloof. No church leader will be perfect, but there does need to be a general consistency in virtue and godliness. We all need grace and forgiveness, and we all need to imitate Christ. It is our hope and prayer that God will fill his children with the Holy Spirit, so that they can follow Jesus effectively and fruitfully. The church of Christ will flourish when godly servant leaders put others before themselves and live to honour and glorify their risen Lord and Saviour. Jesus, the head of the church, is the ultimate servant leader. Since he is our example, servant leadership must be our goal.

## REFLECTION QUESTIONS

1. How has your culture influenced your church's view of church leadership?

2. If a leader in a church is not acting as a servant leader, what can be done to correct the situation?

3. What other biblical passages are relevant to the topic of church leadership?

4. This chapter focused on the moral virtues required for servant leadership. Much more could be said about the type of strengths that leaders need to exhibit. What kinds of strengths do leaders need? How can a leader balance serving others with leading them?

## Also available from Carey Printing Press

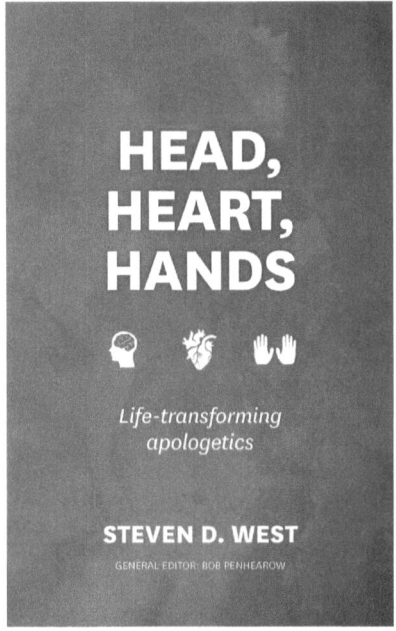

A Christian's entire life should proclaim and defend the reality of the gospel of Jesus Christ. This book demonstrates that the Christian faith is rationally sufficient (head), and emotionally and morally satisfying (heart), and that the seed of the gospel generates practical acts of loving service in the world (hands). The best defence of the Christian faith is one that engages every area of life and every component of our human nature. Everything is transformed by the truth and power of Jesus, and this book helps Christians see how their renewed minds, cleansed hearts and sanctified hands are all necessary in the defence of their faith.

One of the special features of this book is that it illustrates its principles by the life, terminal cancer and death of a young Christian husband and father. Andrew Rozalowsky was a very intelligent young man who was training to be a New Testament scholar. He was married to a wonderful woman, and they had two beautiful young sons. As his cancer took its toll, Andrew demonstrated the power and persuasiveness of a hope that is in Christ alone. Andrew's story is woven throughout the arguments of this book, serving as a practical illustration that the truth and power of the gospel is enough for both life and death.

*Head, Heart, Hands: Life-transforming apologetics*
By Steven D. West
ISBN 978-0-9876841-3-4

## Also available from Carey Printing Press

Using Charles Haddon Spurgeon as a model, this book looks at how the personal spirituality and piety of a pastor is tied to his success and faithfulness in ministry. A Puritan and Calvinistic Baptist heritage served to mould Spurgeon's life and the development of his ministry. The pivotal influence of the writings of John Bunyan, John Gill and Andrew Fuller are examined in detail.

Spurgeon's faithful and intimate walk with God undergirded his preaching, teaching and writing ministries and provided the impetus that led him to establish many organizations and societies to relieve poverty, assist people with addictions and provide homes for orphans.

C.H. Spurgeon was the preeminent Baptist evangelist/preacher in nineteenth-century England. His sermons crossed the Atlantic and were printed and distributed each week for many years. Today, the advice he gave to ministry students, his extensive sermon collection and his written works continue to teach.

*Exploring C.H. Spurgeon's Key to Ministerial Success*
By Bob Penhearow
ISBN 978-0-9876841-0-3

## Also available from Carey Printing Press

In this first volume of a systematic trilogy, the reader is confronted with the gracious character of God and the glorious being of God in Trinity. The wonderful revelation of God in and through his Word and creative acts are then examined, followed by the immutable sovereign decrees of God and God's wisdom through providence.

To help with our understanding of biblical truth, Dr. Bayes has examined the formation of doctrine from the early church creeds, such as the Apostles' Creed and the Nicene Creed, to modern creeds and confessions.

Each chapter closes with an emphasis on pastoral application, reinforcing the fact that theology is meant to practically equip the believer for both life and worship.

*Systematic Theology 1: Systematics for God's Glory*
*God, creation, decrees and providence*
By Jonathan Bayes
ISBN 978-0-9876841-2-7 (paperback)
ISBN 978-0-9876841-1-0 (hardcover)

www.ingramcontent.com/pod-product-compliance
Lightning Source LLC
Chambersburg PA
CBHW020225170426
43201CB00007B/315